Designers' handbook to Eurocode 2

Part 1.1: Design of concrete structures

A. W. Beeby and R. S. Narayanan

┌┐
┴┘ Thomas Telford, London

Published by Thomas Telford Services Ltd, Thomas Telford House, 1 Heron Quay, London E14 4JD

First published 1995

Distributors for Thomas Telford books are
USA: American Society of Civil Engineers, Publications Sales Department, 345 East 47th Street, New York, NY 10017-2398
Japan: Maruzen Co Ltd, Book Department, 3–10 Nihonbashi 2-chome, Chuo-ku, Tokyo 103
Australia: DA Books & Journals, 648 Whitehorse Road, Mitcham 3132, Victoria

A catalogue record for this book is available from the British Library

Classification
Availability: Unrestricted
Content: Guidance based on research and best current practice
Status: Refereed
User: Civil and structural engineering designers

ISBN: 0 7277 1668 9

© A. W. Beeby and R. S. Narayanan, 1995

All rights, including translation reserved. Except for fair copying, no part of this publication may be reproduced, stored in a retrieval system or transmitted in any form or by any means, electronic, mechanical, photocopying or otherwise, without the prior written permission of the Publications Manager, Publications Division, Thomas Telford Services Ltd, Thomas Telford House, 1 Heron Quay, London E14 4JD.

The book is published on the understanding that the authors are solely responsible for the statements made and opinions expressed in it and that its publication does not necessarily imply that such statements and or opinions are or reflect the views or opinions of the publishers.

Every effort has been made to ensure that the statements made and the opinions expressed in this publication provide a safe and accurate guide; however, no liability or responsibility of any kind can be accepted in this respect by the publishers or the authors.

Typeset in Great Britain by MHL Typesetting Ltd, Coventry

Printed and bound in Great Britain by Redwood Books, Trowbridge, Wiltshire

Contents

CHAPTER 1. INTRODUCTION

1.1. Scope   1
1.2. Layout   1
1.3. Terminology   2
1.4. Related documents   2

CHAPTER 2. BASIS OF DESIGN

2.1. Notation   3
2.2. General   3
2.3. Fundamental requirements   3
2.4. Limit states   4
2.5. Actions   4
    Classifications, 4
    Characteristic values of action, 4
    Design values of action, 5
    Simplified load combinations, 9
2.6. Material properties   10
    Characteristic values, 10
    Design values, 10
2.7. Geometric data   10
2.8. Verification   10
    Ultimate limit state, 10
    Serviceability limit state, 11
2.9. Durability   11

CHAPTER 3. ANALYSIS

3.1. Introduction   17
3.2. Load cases and combinations   17
3.3. Imperfections   21
    General, 21
    Global analysis, 22
    Design of slender elements, 23
    Members transferring forces to bracing elements, 23

CONTENTS

- 3.4. Second-order effects — 23
- 3.5. Time-dependent effects — 24
- 3.6. Design by testing — 24
- 3.7. Structural analysis — 24
  - Elastic analysis with or without redistribution, 24
  - Plastic analysis, 29
  - Non-linear analysis, 31
  - Strut-and-tie models, 32
- 3.8. Design aids and simplifications — 34
  - One-way spanning slabs and continuous beams, 34
  - Two-way spanning slabs, 36
  - Flat slabs: methods of analysis, 40
  - Beams, 45
  - Simplifications, 45

CHAPTER 4. MATERIALS AND DESIGN DATA

- 4.1. Concrete — 47
  - General, 47
  - Strength classes and compressive strength f_{ck}, 47
  - Tensile strength f_{ct}, 47
  - Shear strength τ_{Rd}, 47
  - Bond strength, 48
  - Deformation properties, 49
- 4.2. Reinforcement — 51
  - General, 51
  - Strength f_{yk}, 52
  - Ductility, 52

CHAPTER 5. DESIGN OF SECTIONS FOR BENDING AND AXIAL FORCE

- 5.1. Basic assumptions — 54
  - Stress–strain curves, 54
  - Assumptions relating to the strains at ultimate, 57
- 5.2. Design equations — 59
  - Singly reinforced beams and slabs, 59
  - Doubly reinforced rectangular sections, 62
  - Design of flanged sections (T- or L-beams), 63
  - Design of rectangular column sections, 64
  - Design for biaxial bending, 72
  - Design of prestressed sections, 87
  - Brittle failure and hyperstrength, 87

CHAPTER 6. SHEAR, PUNCHING SHEAR AND TORSION

- 6.1. Shear: general — 88
- 6.2. Background to the EC2 provisions — 88
  - Members without shear reinforcement, 88
  - Members with shear reinforcement, 89
  - Maximum shear strength of a section, 91
  - Shear capacity enhancement near supports, 92
  - Summary, 94

6.3.	Summary of the provisions in *clause 4.3.2*	94
6.4.	Punching shear	99
	General, *99*	
	Critical perimeter, *99*	
	Design shear force, *101*	
	Shear resistance of slabs without shear reinforcement, *103*	
	Reinforcement for punching shear, *103*	
6.5.	Torsion	106
	Introduction, *106*	
	Evaluation of torsional moments, *106*	
	Design verification, *106*	
	Combined torsion and bending, *108*	
	Combined torsion and shear, *109*	
	Detailing, *109*	

CHAPTER 7. SLENDER COLUMNS AND BEAMS

7.1.	Scope	111
7.2.	Background to design of columns for slenderness effects	111
7.3.	Design for slenderness effects	116
	Assumptions for non-linear analysis and outline procedure, *117*	
	Simplified method, *119*	
	Other factors, *125*	
	Simplified method for the design of sway frames, *127*	
	Walls, *130*	
	Lateral buckling of slender beams, *130*	

CHAPTER 8. SERVICEABILITY

8.1.	General	132
	Assessment of design action effects, *132*	
	Material properties, *133*	
8.2.	Limitation of stresses under serviceability conditions	135
	General, *135*	
	Procedure for stress checks, *138*	
8.3.	Control of cracking	156
	General, *156*	
	Principles of the cracking phenomena, *157*	
	Derivation of crack prediction formulae, *159*	
	Minimum areas of reinforcement, *162*	
	Checking cracking without direct calculation, *166*	
	Checking cracking by direct calculation, *168*	
8.4.	Control of deflections	169
	General, *169*	
	Deflection limits, *169*	
	Design loads, *170*	
	Material properties, *171*	
	Model of behaviour, *171*	
	Simplified approach to checking deflections, *181*	

CHAPTER 9. DURABILITY

- 9.1. General — 184
- 9.2. Background — 184
  - Historical perspective, *184*
  - Common mechanisms leading to the deterioration of concrete structures, *185*
  - Relative importance of deterioration mechanisms, *187*
- 9.3. Design for durability — 188
  - General, *188*
  - Definition of aggressivity of the environment, *188*
  - Measures to resist environmental aggressivity, *189*

CHAPTER 10. DETAILING

- 10.1. General — 191
- 10.2. Discussion of the general requirements — 191
  - Cover to bar reinforcement, *192*
  - Spacing of bars, *192*
  - Minimum diameters of bends, *192*
  - Bond, *192*
  - Anchorage, *194*
  - Splices (laps), *195*
  - Bars with $\phi > 32$ mm, *198*
  - Welded mesh, *199*
  - Welded mesh using smooth wires, *201*
  - Beams, *202*
  - Slabs, *206*
  - Columns, *210*
  - Walls, *211*
  - Corbels, *213*
  - Nibs, *214*
  - Shear reinforcement in flat slabs, *214*
  - Bundled bars, *215*

CHAPTER 11. PRESTRESSED CONCRETE

- 11.1. General — 216
- 11.2. Summary of main clauses — 216
- 11.3. Durability — 216
- 11.4. Design data — 218
  - Concrete, *218*
  - Prestressing steel, *219*
- 11.5. Design of sections for flexure and axial load — 220
  - Ultimate limit state, *220*
  - Serviceability limit state, *221*
- 11.6. Design of sections for shear and torsion — 224
  - Shear, *224*
  - Torsion, *227*

11.7.	Prestress losses	228
	General, *228*	
	Friction in jack and anchorages, *228*	
	Duct friction, *228*	
	Elastic deformation, *228*	
	Anchorage draw-in or slip, *228*	
	Time-dependent losses, *228*	
11.8.	Anchorage zones	229
	Pretensioned members, *229*	
	Post-tensioned members, *230*	
11.9	Detailing	230
	Specing of tendons and ducts, *230*	
	Anchorages and couplers, *231*	
	Minimum area of tendons, *231*	
	Tendon profiles, *232*	
	Shear reinforcement, *232*	

APPENDIX: DESIGN OF SWAY FRAMES IN EC2

A.1.	Development of a simplified method	233
	Concluding discussion, *241*	
	Assumptions, *241*	
	Theory, *241*	
	Rules, *241*	

CHAPTER 1

Introduction

1.1. Scope
Eurocode 2, Design of Concrete Structures, will apply to the design of building and civil engineering structures in plain, reinforced and prestressed concrete. Eurocode 2 is written in several parts: this manual is concerned with *Part 1: General rules and rules for buildings*. Eurocode 2: Part 1 (the formal title is DD ENV 1992-1-1:1992, but it is referred to as EC2 or ENV EC2 throughout this manual), is written in such a way that its principles will generally apply to all the parts when they are subsequently developed. The specific rules have been worked out for building structures only. It is envisaged that the further parts will complement and adapt Part 1 as appropriate, and develop rules as appropriate for the particular type of structure considered.

EC2 has seven chapters and four appendices. The main chapters deal with matters required for day-to-day design; the appendices give further guidance that will occasionally be required for special situations.

Compliance with EC2 will satisfy the requirements of the Construction Products Directive in respect of mechanical resistance.

1.2. Layout
The EC2 clauses are set out as principles and application rules. Principles are distinguished by the prefix 'P'. Application rules are indented. Principles are general statements, and definitions for which there are no alternatives. Application rules are generally accepted methods, which follow the principles and satisfy their requirements. Alternative rules are permissible provided that it can be demonstrated that they comply with the principles and that equivalent resistance, serviceability and durability will be achieved. Here equivalence refers to the reliability of results. Thus, procedures in the current national codes are by and large likely to be acceptable, as the principles are likely to be similar. For example, tying requirements in ENV EC2 are in general terms, whereas the UK Code gives specific provisions, which will comply with the requirements of EC2.

Some numerical values are given within a box (☐), and are referred to as boxed values. These are given only as indications. Each member state is required to fix the boxed values applicable within its jurisdiction. The boxed values will be found in the National Application Documents (NAD) which will be produced by each country. In the long run it is hoped that many of these values can be harmonized across the European Community. The manual uses the indicative values throughout.

Chapters are arranged generally by reference to phenomena, rather than to the type of element as in UK codes. For example, there are chapters on bending, shear, buckling, etc., but not on beams, slabs or columns. However, there are some exceptions, such as clauses referring to deep beams and corbels. Such a layout is more efficient, as considerable duplication is avoided. It also promotes a better understanding of structural behaviour.

1.3. Terminology

Generally, the language and terminology employed will be familiar to most engineers. In any case, these definitions are generally provided at the beginning of each chapter of EC2. A few commonly occurring features are noted below.

Loads are generally referred to as 'actions', to characterize the generalized format of the code. Actions refer not only to the forces directly applied to the structure but also to imposed deformations, such as temperature effect or settlement. These are referred to as 'indirect actions' and the suffix 'IND' is used to identify them.

Actions are further subdivided as permanent (G) (dead loads), variable (Q) (live loads or wind loads) and accidental (A). Prestressing (P) is treated as a permanent action in most situations.

Action effects are the internal forces, stresses, bending moments, shear and deformations caused by actions.

Characteristic values of any parameter are distinguished by a suffix 'k'. Design values carry a suffix 'd' and take account of partial safety factors.

All the formulae and expressions in EC2 are in terms of the cylinder strength of concrete, denoted by f_{ck}. The relationship between the cylinder and cube strengths is set out within EC2. Generally, the strength class is referred to as 'C(cylinder strength)/(cube strength)', for example C25/30 refers to cylinder strength of 25 N/mm^2 and a corresponding cube strength of 30N/mm^2. It is not anticipated that the quality control procedures for the production of concrete will switch to cylinder strength where the cube strength is now used (e.g. as in the UK).

1.4. Related documents

EC2 refers to international standards where relevant standards are available. Other standards needed to make this Eurocode fully operational are under preparation by CEN (Comité Européen de Normalisation—Euro Standards Body). The most important of such guidance documents will be those that give (*a*) values of variable action, (*b*) design for fire, (*c*) properties of reinforcement and prestressing steels.

Each member state is required to prepare a National Application Document (NAD) which will fix the boxed values and also provide additional guidance to make the ENV EC2 usable. For instance, EC2 gives a ductility classification of reinforcement. NAD is required to advise how locally available reinforcement is to be regarded in this classification. Relevant CEN standards/codes are not likely to be available when the Eurocode is published as an ENV document (pre-standard). The NAD may recommend the use of the national or ISO codes as an interim measure pending the publication of appropriate CEN standards. The NAD is therefore an essential reference document.

EC2 relies on ENV 206 (*Concrete performance, production, placing and compliance criteria*) for the specification of concrete mixes to ensure durability in various exposure conditions, which are defined in the same document.

CHAPTER 2

Basis of design

2.1. Notation
In this manual, symbols are defined locally where they occur. The following is a list of symbols that occur throughout the manual.

A_k	characteristic value of an accidental action
$G_{k,\text{inf}}$	lower characteristic value of a permanent action
$G_{k,j}$	characteristic value of the permanent action j
$G_{k,\text{sup}}$	upper characteristic value of a permanent action
P_k	characteristic value of prestressing force
$Q_{k,i}$	characteristic value of the variable action i
X_k	characteristic value of a material property
$\gamma_{Ga,j}$	partial safety factor for permanent action j for accidental design situations
$\gamma_{G,j}$	partial safety factor for permanent action j for persistent and transient design situations
γ_m	partial safety factor for the material property
γ_p	partial safety factor for prestressing force
$\gamma_{Q,i}$	partial safety factor for variable action i
ψ_0, ψ_1, ψ_2	multipliers for the characteristic values of variable actions to produce combination, frequent and quasi-permanent values respectively of variable actions to be used in various verifications

2.2. General
Most of this chapter* is material-independent and as such is common to all Eurocodes. It is written in a generalized format. For those who are used to short prescriptive codes, this chapter may seem over-long and all-embracing. However, the principles are relatively simple to follow. Familiarity should quickly dispel any initial reservations.

EC2 is drafted in accordance with limit state principles.

2.3. Fundamental requirements
Four basic requirements can be summarized as follows. The structure should

(a) have an acceptably low probability of becoming unfit for its intended use
(b) sustain all the loads likely to occur during construction and use with appropriate reliability
(c) have adequate durability in relation to maintenance costs

It is anticipated that the chapter on Basis of design will be largely transformed to a separate part of EC1

(d) not be damaged by accidents to an extent disproportionate to the original cause.

EC2 does not specify the 'acceptable probability' or 'appropriate reliability', therefore all that can be said is that the Committee considers that the safety level implied by the use of the partial safety factors given in EC2 is adequate. No attempt appears to have been made to check the 'safety levels' across materials; therefore the 'degree of safety' is likely to be different in structures built in different materials (e.g. structural steel and reinforced concrete).

2.4. Limit states

Limit states are defined as states beyond which the structure infringes an agreed performance criterion. Two basic groups of limit states to be considered are (a) ultimate limit states and (b) serviceability limit states.

Ultimate limit states are those associated with collapse or failure, and generally govern the strength of the structure or components. They also include loss of equilibrium or stability of the structure as a whole. As the structure will undergo severe deformation prior to reaching collapse conditions (e.g. beams becoming catenaries), for simplicity these states are also regarded as ultimate limit states, i.e. a condition between serviceability and ultimate limit states. These are equivalent to collapse, as they will necessitate replacement of the structure or element.

Serviceability limit states generally correspond to conditions of the structure in use. They include deformation, cracking and vibration which

(a) damage the structure or non-structural elements (such as finishes, partitions, etc.) or the contents of buildings (such as machinery)
(b) cause discomfort to the occupants of buildings
(c) adversely affect appearance, durability or watertightness and weather-tightness.

They will generally govern the stiffness of the structure and the detailing of reinforcement within it.

Figure 2.1 illustrates a typical load—deformation relationship of a reinforced concrete structure and the limit states.

2.5. Actions

2.5.1. Classifications

An action is a direct force (load) applied to a structure or an imposed deformation, such as settlement or temperature effects. The latter is referred to as an indirect action. Accidental actions are caused by unintended events that are generally of short duration and have a very low probability of occurrence.

The main classification of actions for common design is given in Table 2.1.

2.5.2. Characteristic values of action

Loads vary in time and space. In limit state design the effects of loads, which are factored suitably, are compared with the resistance of the structure, which is calculated by suitably discounting the material properties. In theory, characteristic values are obtained statistically from existing data. In practice, however, this is very rarely possible, particularly for imposed loads whose nominal values, often specified by the client, are used as characteristic loads. In countries

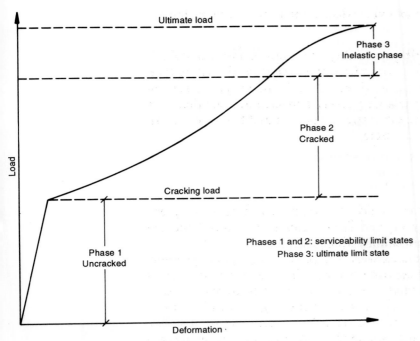

Fig. 2.1. Typical load–deformation of reinforced concrete structure: phases 1 and 2 = serviceability limit states; phase 3 = ultimate limit state

Table 2.1. Classification of actions

Permanent action	Variable action	Accidental action
(a) Self-weight of structures, fittings and fixed equipment	(a) Imposed floor loads	(a) Explosions
(b) Prestressing force	(b) Snow loads	(b) Fire
(c) Water and earth loads	(c) Wind loads	(c) Impact from vehicles
(d) Indirect action, e.g. settlement of supports	(d) Indirect action, e.g. temperature effects	

where wind and snow data have been gathered over a period, it will be possible to prescribe a statistically estimated characteristic value.

Characteristic values for loads will be given in EC1 (*Eurocode for actions on structures*).

For permanent actions that vary very little about their mean (such as weights of materials), the characteristic value corresponds to the mean value. When the variation is likely to be large (e.g. for walls or slabs cast against earth surface with random variations in thickness, or loads imposed by soil fill), upper and lower characteristic values (corresponding ideally to 95 and 5 percentiles) will need to be assessed. These values will apply during the life of the structure, and are denoted by $G_{k,sup}$ and $G_{k,inf}$ respectively.

2.5.3. Design values of action

The values of actions to be used in design are governed by a number of factors, including the following.

(a) The nature of the load. Whether the action is permanent, variable or accidental, as the confidence in the description of each will vary.

(b) The limit state being considered. Clearly, the design value must be higher for ultimate limit state than for serviceability. Further, in serviceability conditions loads vary with time, and the design load to be considered could vary substantially. Realistic serviceability loads should be modelled depending on the aspect of the behaviour being checked (e.g. deflection, cracking, settlement). For example, creep and settlement are functions of permanent loads only.

(c) The number of variable loads acting simultaneously. Statistically it is improbable that all loads will act at their full characteristic value at the same time. To allow for this, the characteristic values of action will need modification.

Consider the case of permanent action G_k and one variable action Q_k only. For the ultimate limit state the characteristic values should be magnified and the load may be represented as $\gamma_G.G_k + \gamma_Q.Q_k$, where γ_G and γ_Q are the partial safety factors. The values of γ_G and γ_Q will be different, and will reflect the fact that the variabilities of the two loads are different. The γ factors account for

(a) the possibility of unfavourable deviation of the loads from the characteristic values
(b) inaccuracies in the analyses
(c) unforeseen stress redistributions
(d) variation in the geometry of the structure and its elements, as it affects the determination of the action effects.

Now consider the case of a structure subject to variable actions Q_1 and Q_2 simultaneously. If Q_1 and Q_2 are independent, i.e. the occurrence and magnitude of Q_1 does not depend on the occurrence and magnitude of Q_2 and vice versa, then it would be unrealistic to use $(\gamma_{Q1}.Q_{k1} + \gamma_{Q2}.Q_{k2})$, as the two loads are unlikely to act at their maximum at the same time. Joint probabilities will need to be considered to ensure that the probability of occurrence of the two loads is the same as that of a single load. It will be more reasonable to consider one load at its maximum in conjunction with a reduced value for the other load. Thus there are two possibilities

$$[\gamma_{Q,1}.Q_{k,1} + \psi_{0,2}(\gamma_{Q,2}.Q_{k,2})]$$
$$[\psi_{0,1}(\gamma_{Q,1}.Q_{k,1}) + \gamma_{Q,2}.Q_{k,2}]$$

Multiplication by ψ_0 is said to produce a combination value of the load. In theory, γ and ψ_0 values can vary with each load. EC1 will provide values of ψ_0 for different loads. In the interim, values are given in the NAD to EC2. See also Table 2.3 below. The method of deriving ψ values is outlined in the addenda to ISO Code 2394:1986. In practice, the designer will not have sufficient information to vary the γ values in most cases. Table 2.2 below gives the γ values recommended by EC2.

The above discussion illustrates the thinking behind the method of combining loads for an ultimate limit check. Similar logic is applied to the estimation of loads for the various serviceability checks.

Some explanation is appropriate to clarify the rather obscurely drafted clauses dealing with static equilibrium, i.e. stability checks [EC2 *Clauses 2.3.2.3P(3) and 2.3.3.1(3)*]. These deal with the design loading caused by a *single* permanent action, in structures that may be sensitive to variations in loads in different parts of the structure.

Clause 2.3.2.3P(3)
Clause 2.3.3.1(3)

For example, consider a simple overhanging cantilever beam. In the cantilever section the self-weight will be destabilizing, whereas it will contribute to stability in the anchor span. The clauses noted above will apply. If some permanent actions are present in the cantilever section only, then these clauses do not apply and the partial safety factors to be used will be those associated with strength checks as shown in Table 2.2.

Note that the factors $\gamma_{G,inf}$ and $\gamma_{G,sup}$, in *clause 2.3.3.1(3)* are different to the factors applying to strength checks. In the overhanging cantilever beam example noted above, the multiplier for self-weight in the cantilever section will be $1 \cdot 1$ ($\gamma_{G,sup}$) and that in the anchor span will be $0 \cdot 9$ ($\gamma_{G,inf}$). The possible explanation for $\gamma_{G,sup}$ being $1 \cdot 1$ and not $1 \cdot 35$ is that

Clause 2.3.3.1(3)

(a) the variability in self-weight of the element is unlikely to be large
(b) the factor $1 \cdot 35$ has built into it an allowance for structural performance (which is necessary only for strength checks)
(c) the loading in the cantilever will also generally include variable actions.

Various generalized combinations of loads can be expressed symbolically as in (a)–(e) below. Note that the '+' symbol in the expressions does not have the normal mathematical meaning, as the directions of loads could be different. It is better to read it as 'combined with'.

2.5.3.1. ULTIMATE LIMIT STATE.

(a) Fundamental combination (persistent and transient situations)

$$\Sigma \gamma_{G,j} . G_{k,j} + \gamma_{Q,1} . Q_{k,1} + \sum_{i>1} \gamma_{Q,i} . \psi_{0,i} . Q_{k,i}$$

The above combination assumes that a number of variable actions are present at the same time. $Q_{k,i}$ is the dominant load if it is obvious, otherwise each load is in turn treated as a dominant load and the others as secondary. The dominant load is then combined with the combination value of the secondary loads. Both are multiplied by their respective γ values.

(b) Accidental design situation

$$\Sigma \gamma_{GA,j} . G_{k,j} + A_d + \psi_{1,1} . Q_{k,1} + \sum_{i>1} \psi_{2,i} . Q_{k,i}$$

where A_d is the design value of accidental action.

Accidents are unintended events such as explosions, fire or vehicular impact, which are of short duration and have a low probability of occurrence. Also, a degree of damage is generally acceptable in the event of an accident. The loading model should attempt to describe the magnitude of other variable loads that are likely to occur in conjunction with the accidental load. Accidents generally occur in structures in use. Therefore the values of variable actions will be less than those used for the fundamental combination of loads in (a) above. To provide a realistic variable load for combination with the accidental load, the variable actions are multiplied by different (and generally lower) ψ factors, i.e. ψ_1 and ψ_2. The multiplier ψ_1 is applied to the dominant action, and ψ_2 to the others. Where the dominant action is not obvious, each variable action present is in turn treated as dominant and multiplied by ψ_1, and is combined with the other loads multiplied by ψ_2. The γ_Q for accidental situations is unity.

Multiplication by ψ_1 is said to produce a frequent value of the load, and multiplication by ψ_2 the quasi-permanent value. Values for ψ_1 and ψ_2 will be given in EC1; in the interim, values are given in the NAD for EC2. See also Table 2.3 below.

2.5.3.2. SERVICEABILITY LIMIT STATE.

(c) Rare combination

$$\Sigma G_{k,j}(+P) + Q_{k,1} + \sum_{i>1} \psi_{0,i} \cdot Q_{k,i}$$

This represents a combination of service loads that can be considered infrequent. It might be appropriate for checking states such as microcracking or possible local non-catastrophic failure of reinforcement leading to large cracks in sections.

(d) Frequent combination

$$\Sigma G_{k,j}(+P) + \psi_{1,1} Q_{k,1} + \sum_{i>1} \psi_{2,i} \cdot Q_{k,i}$$

This represents a combination likely to occur relatively frequently in service conditions, and is used for checking some aspects of cracking.

(e) Quasi-permanent combination

$$\Sigma G_{k,j}(+P) + \Sigma \psi_{2,i} \cdot Q_{k,i}$$

This will provide an estimate of sustained loads on the structure and will be suitable for checking creep, settlement, etc.

It should be realized that the above combinations describe the magnitude of loads that are likely to be present simultaneously. The actual arrangement of loads in position and direction within the structure to create the most critical effect is a matter of structural analysis (e.g. loading alternate or adjacent spans in continuous beams).

γ and ψ values are given in Tables 2.2 and 2.3.

Table 2.2. Partial safety factors for actions in building structures: ultimate limit state

Action	Combination	
	Fundamental	Accidental
Permanent actions caused by structural and non-structural components		
Stability check		
Unfavourable	1·10	1·00
Favourable	0·90	1·00
Other checks		
Unfavourable	1·35	1·00
Favourable	1·00	1·00
Variable actions		
Unfavourable	1·50	1·00
Accidental actions	—	1·00

Values apply to persistent and transient design situations.
For accidental design situations $\gamma_{G,A} = 1·0$.
Partial safety factor for prestressing force γ_p is generally 1·0.
For imposed deformation, γ_Q for unfavourable effects is 1·2 when linear methods are used. For non-linear methods, $\gamma_Q = 1·5$.

BASIS OF DESIGN

Table 2.3. ψ values

Variable actions	ψ_0	ψ_1	ψ_2
Imposed loads			
Dwellings	0.5	0.4	0.2
Offices and stores	0.7	0.6	0.3
Parking	0.7	0.7	0.6
Wind loads	0.7	0.2	0
Snow loads	0.7	0.2	0

For the purposes of EC2 these three categories of variable actions should be treated as separate and independent actions.

Examples 2.1–2.4 below (similar examples were prepared for another publication commissioned by the Department of the Environment (UK)) illustrate the use of the combinations noted above. However, in practice simplified methods given in section 2.5.4 below are likely to be all that is needed for most structures. Also, in practical examples the dominant loads are likely to be fairly obvious, and therefore the designer will generally not be required to go through all the combinations.

2.5.4. Simplified load combinations

EC2 permits the following simplified combinations as an alternative to the combinations given in section 2.5.3 above.

(a) Design situations with one variable action only

$$\Sigma \gamma_{G,j} \cdot G_{k,j} + 1 \cdot 5 \, Q_{k,1} \qquad \text{Ultimate limit state}$$

$$\Sigma G_{k,j} \, (+P) + Q_{k,1} \qquad \text{Serviceability limit state}$$

(b) Design situation with two or more variable actions

$$\Sigma \gamma_{G,j} \, G_{k,j} + 1 \cdot 35 \sum_{i \geq 1} Q_{k,i} \qquad \text{Ultimate limit state}$$

$$\Sigma G_{k,j} \, (+P) + 0 \cdot 9 \sum_{i \geq 1} Q_{k,i} \qquad \text{Serviceability limit state}$$

For each Q_i that can occur independently, combination (a) above should also be considered.

For normal building structures the above expressions for ultimate limit state may also be represented as shown in Table 2.4.

Table 2.4. Partial safety factors for simplified load combinations: ultimate limit state

Load combination	Permanent load		Variable load			Prestress
	Adverse	Beneficial	Imposed		Wind	
			Adverse	Beneficial		
Permanent + imposed	1.35	1.00	1.50	0	—	1.00
Permanent + wind	1.35	1.00	—	—	1.50	1.00
Permanent + imposed + wind	1.35	1.00	1.35	0	1.35	1.00

2.6. Material properties

2.6.1. Characteristic values
A material property is represented by a characteristic value X_k which in general corresponds to a fractile (commonly 5%) in the statistical distribution of the property, i.e. it is the value below which the chosen percentage of all test results are expected to fall.

Generally, in design, only one (lower) characteristic value will be of interest. However, in some problems such as cracking in concrete, an upper characteristic value may be required, i.e. the value of the property (such as the tensile strength of concrete) above which only a chosen percentage of the values are expected to fall.

2.6.2. Design values
In order to account for the differences between the strength of test specimens of the structural materials and their strength in situ, the strength properties will need to be reduced. This is achieved by dividing the characteristic values by partial safety factors for materials γ_m. Thus, the design value $X_d = X_k/\gamma_m$. Although this is not stated in EC2, γ_m also accounts for local weaknesses and inaccuracies in the assessment of resistance of the section.

The values of partial safety factors for material properties are given in Table 2.5.

2.7. Geometric data
The structure is normally described using the nominal values for the geometrical parameters. Variability of these parameters is generally negligible compared to the variability associated with the values of actions and material properties.

In special problems such as buckling and global analyses, geometrical imperfections should be taken into account. EC2 specifies values for these in the relevant sections. Traditionally, geometrical parameters are modified by additive factors.

2.8. Verification

2.8.1. Ultimate limit state
(a) When considering overall stability, it should be verified that the design effects of destabilizing actions are less than the design effects of stabilizing actions.
(b) When considering rupture or excessive deformation of a section, member

Table 2.5. Partial safety factor for material properties

Combination	Concrete γ_c	Reinforcement and prestressing tendons γ_s
Fundamental	1·5	1·15
Accidental, except earthquakes	1·3	1·0

These values apply to ultimate limit state. For serviceability, $\gamma_m = 1$. These values apply if the quality control procedures stipulated in EC2 are followed. Different values of γ_c may be used if justified by commensurate control procedures. These values do not apply to fatigue verification.

or connection it should be verified that the design value of internal force or moment is less than the design value of resistance.

(c) It should be ensured that the structure is not transformed into a mechanism unless actions exceed their design values.

2.8.2. Serviceability limit state

(d) It should be verified that the design effects of actions do not exceed a nominal value or a function of certain design properties of materials, for example deflection under quasi-permanent loads should be less than span/250 or compression stress under rare combination of loads should not exceed $0.6f_{ck}$.

In most cases detailed calculations using various load combinations are unnecessary, as EC2 stipulates simple compliance rules.

2.9. Durability

As one of the fundamental aims of design is to produce a durable structure, a number of matters will need to be considered early in the design process. These include:

- the use of the structure
- the required performance criteria
- the expected environmental conditions
- the composition, properties and performance of the materials
- the shape of members and the structural detailing
- the quality of workmanship, and level of control
- the particular protective measures
- the maintenance during the intended life.

The environmental conditions should be considered at the design stage to assess their significance in relation to durability and to enable adequate provisions to be made for protection of the materials.

Example 2.1
For the frame shown in Fig. 2.2, identify the various load combinations, to check the overall stability. Assume office use for this building. (Note that the load combinations for the design of elements could be different.)

(i) Notation
G_{kR} characteristic dead load/m   (roof)
G_{kF} characteristic dead load/m   (floor)
Q_{kR} characteristic live load/m   (roof)
Q_{kF} characteristic live load/m   (floor)
W_k characteristic wind load/frame at each floor level

(ii) Fundamental load combination to be used is

$$\sum \gamma_{G,j} \cdot G_{k,j} + \gamma_{Q,1} + \sum_{i>1} \gamma_{Q,i} \cdot \psi_{0,i} \cdot Q_{k,i}$$

Clause 2.3.2.2P(2)

As the stability will be sensitive to possible variation of dead loads, it will be necessary to allow for this as per *clause 2.3.2.3P(3)*. Take

Clause 2.3.2.3P(3)

$\gamma_{G,\text{inf}} = 0.9$
$\gamma_{G,\text{sup}} = 1.35$
$\gamma_Q = 1.5$
ψ_0 — imposed loads (offices) = 0.7
ψu — wind loads = 0.7

See Table 2.2 and NAD Table 1.

Case 1
Treat the wind load as the dominant load (Fig. 2.3).

Case 2
Treat the imposed load on the roof as the dominant load (Fig. 2.4).

Case 3
Treat the imposed load on the floors as the dominant load (Fig. 2.5).

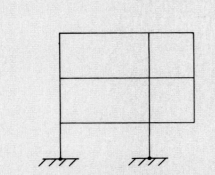

Fig. 2.2. Frame (Example 2.1)

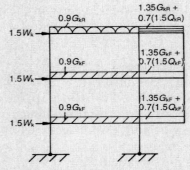

Fig. 2.3. Frame (Example 2.1, Case 1)

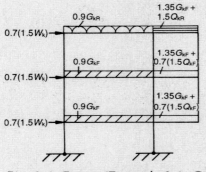

Fig. 2.4. Frame (Example 2.1, Case 2)

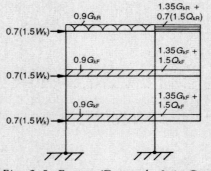

Fig. 2.5. Frame (Example 2.1, Case 3)

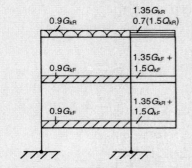

Fig. 2.6. Frame (Example 2.1, Case 4)

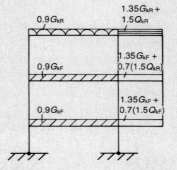

Fig. 2.7. Frame (Example 2.1, Case 5)

Case 4
Consider the case without wind loading, treating imposed floor loads as the dominant load (Fig. 2.6).

Case 5
Consider the case without wind loading, treating the imposed roof loads as the dominant loads (Fig. 2.7).
Note: When the wind loading is reversed, another set of combinations will need to be considered. However, in problems of this type the designer is likely to arrive at the critical combinations intuitively rather than by searching through all the theoretical possibilities.

Example 2.2
Identify the various load combinations for the design of the four-span continuous beam for the ultimate limit state. Assume that spans 1–2 and 2–3 are subject to domestic use and spans 3–4 and 4–5 are subject to parking use (Fig. 2.8).

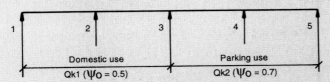

Fig. 2.8. Four-span continuous beam (Example 2.2)

(i) Notation
  G_k characteristic dead load/m
  Q_{k1} characteristic imposed load/m (domestic use)
  Q_{k2} characteristic imposed load/m (parking use)

(ii) Fundamental load combination to be used is

Clause 2.3.2.2P(2)

$$\Sigma \gamma_{G,j} \cdot G_{k,j} + \gamma_{Q,1} Q_{k,1} + \sum_{i>1} \gamma_{Q,i} \cdot \psi_{0,i} \cdot Q_{k,i}$$

For beams without cantilevers, the same value of self-weight can be applied to all spans, i.e. $1\cdot35 G_k$. The load cases to be considered are alternate spans loaded and adjacent spans loaded (Fig. 2.9).

Clause 2.3.2.3(4)
Clause 2.5.1.2(4)

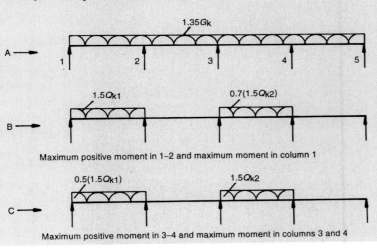

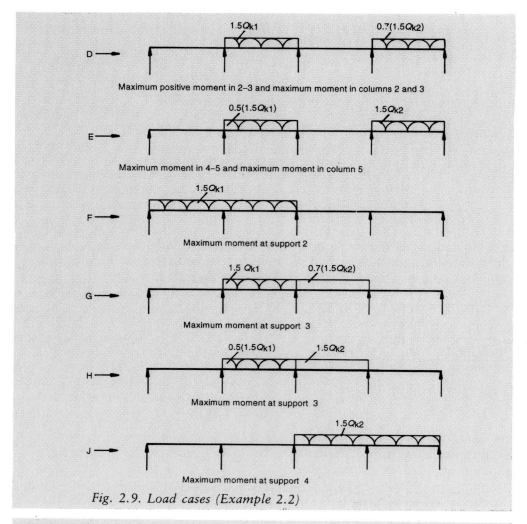

Fig. 2.9. Load cases (Example 2.2)

Example 2.3

For the continuous beam shown in Fig. 2.10, identify the critical load combinations for the design for ultimate limit state. Assume that the beam is subject to dead and imposed loads and a point load at the end of the cantilever arising from dead loads of the external wall.

(i) Notation
- G_k characteristic dead load/m
- Q_k characteristic imposed load/m
- P characteristic point load (dead)

(ii) The fundamental combinations given in *clause 2.3.2.2* should be used (Figs 2.11–2.14). Note that the presence of the cantilever prohibits the use of the same design values of dead loads in all spans. See section 3.2.4 of the manual for possible simplification. The load cases to be considered are as set out in *clause 2.5.1.2(4)*.

Clause 2.3.2.3(4)

Section 3.2.4

Clause 2.5.1.2(4)

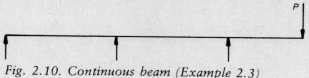

Fig. 2.10. Continuous beam (Example 2.3)

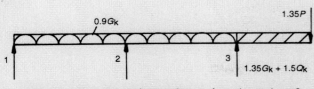

Maximum cantilever BM, maximum anchorage of negative steel over 3; also maximum column moment at 3 (see Fig. 2.14)

Fig. 2.11. Continuous beam (Example 2.3, Case 1)

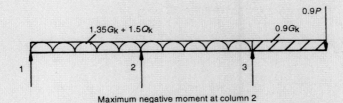

Maximum negative moment at column 2

Fig. 2.12. Continuous beam (Example 2.3, Case 2)

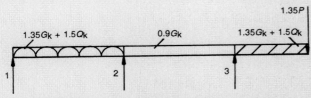

Maximum positive moment 1–2, maximum moment column 1 and possible maximum moment column 2 (see Fig. 2.14)

Fig. 2.13. Continuous beam (Example 2.3, Case 3)

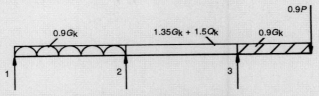

Maximum positive moment 2–3, maximum column moment at 2 (see Fig. 2.13) and maximum column moment at 3 (see columns in Fig. 2.11)

Fig. 2.14. Continuous beam (Example 2.3, Case 4)

Example 2.4

A water tank of depth H m has an operating depth of water h m (Fig. 2.15). Calculate the design lateral loads for ultimate limit state.

According to the EC1 draft, EC1 earth loads are permanent loads (EC1 clause 3.4.8). The same reasoning can be applied to lateral pressures caused by water. NAD for EC2 confirms this (NAD clause 2.3.3.1).

Design can therefore be based on the pressure diagram shown in Fig. 2.16.

Consideration should also be given to the worst credible water load, which in this case will correspond to a depth of H m of water, i.e. water up to the top of the tank.

EC2 permits the variation of the partial safety factor $\gamma_{G,j}$ depending on the knowledge of the load $G_{k,j}$.

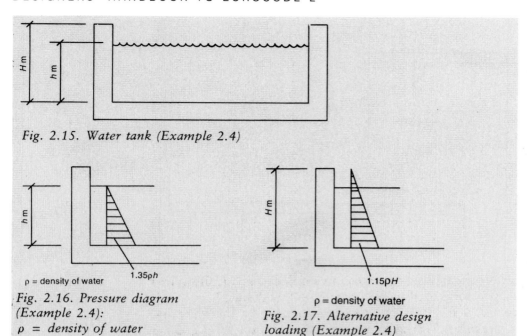

Fig. 2.15. Water tank (Example 2.4)

Fig. 2.16. Pressure diagram (Example 2.4): ρ = density of water

Fig. 2.17. Alternative design loading (Example 2.4)

However, the method of establishing $\gamma_{G,j}$ is not given. The basis adopted in the UK Code BS8110:Part 2 may be adopted, and a factor of 1·15 may be applied in lieu of 1·35. Therefore the alternative design loading will be as shown in Fig. 2.17.

CHAPTER 3

Analysis

3.1. Introduction

The purpose of analysis is the verification of overall stability and establishment of action effects, i.e. the distribution of internal forces and moments. In turn this will enable the calculation of stresses, strains, curvature, rotation and displacements. In certain complex structures the type of analysis used (e.g. finite element analysis) will yield internal stresses and strains and displacements directly.

To carry out the analysis, both the geometry and the behaviour of the structure will need to be idealized.

Commonly, the structure is idealized by considering it as made up of elements depicted in Fig. 3.1.

In terms of the behaviour of the structure, the following methods may be used.

- Elastic analysis
- Elastic analysis with limited redistribution
- Plastic analysis
- Non-linear analysis

Clause 2.5.2.1.(2)–(4)
Clause 2.5.1.1

The first two models are common for slabs and frames; plastic analysis is popular in the design of slabs; non-linear analysis is very rarely used in day-to-day design. The above methods, with the exception of plastic analysis, are suitable for both serviceability and ultimate limit states. Plastic methods can be used only for ultimate limit state (i.e. strength) design.

In addition to global analysis, local analyses may be necessary, particularly when the assumption of linear strain distribution does not apply. Examples of this include

- anchorage zones
- members with significant changes in cross-section including the vicinity of large holes
- beam/column junctions
- locations adjacent to concreted loads.

In these cases strut-and-tie methods (a plastic method) are commonly employed to analyse the structures.

3.2. Load cases and combination

In the analysis of the structure, the designer should consider the effects of the realistic combinations of permanent and variable actions. Within each set of combinations (e.g. dead and imposed loads) a number of different arrangements

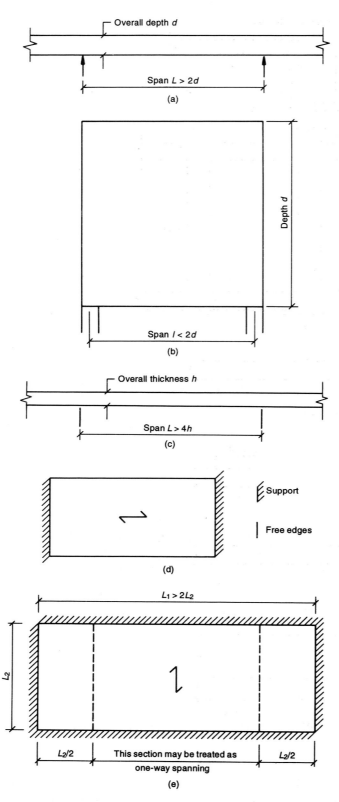

Fig. 3.1. Definition of structural elements for analysis: (a) beam; (b) deep beam; (c) slab; (d), (e) one-way spanning slab (subject predominantly to ultimate design load); (f) ribbed and waffle slabs (conditions to be met to allow analysis as solid slabs); (g) column; (h) wall

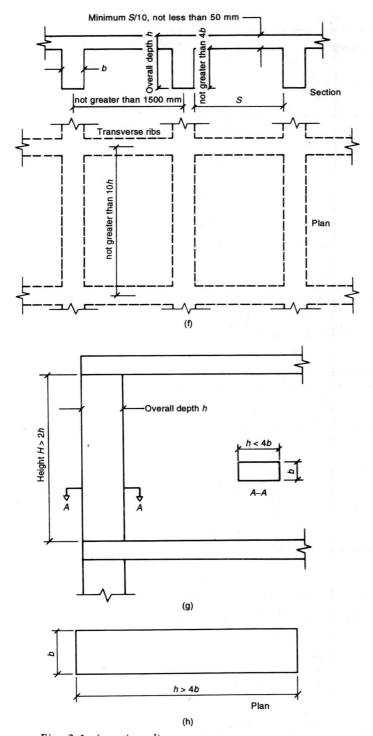

Fig. 3.1. (continued)

of loads (load cases) throughout the structure (e.g. alternate spans loaded and adjacent spans loaded) will need consideration in order to identify an envelope of action effects (e.g. bending moment and shear envelopes) to be used in the design of sections.

Chapter 2 provides the magnitude of the design loads to be used when loads are combined. Account is taken of the probability of loads acting together, and values are specified accordingly.

The Eurocode for actions, EC1, will specify the densities of materials (to enable the calculation of permanent actions and surcharges), and values of variable action (such as imposed gravity, wind and snow loads). It will also provide information for estimation of fire loads in buildings, to enable fire engineering calculations to be carried out.

Until EC1 is finalized, national loading codes will need to be used in conjunction with EC2 (see NAD, clause 4). In the UK these are BS648:1964, Weights of building materials; BS6399:Part 1:1984, Design loading for buildings: Code of Practice for dead and imposed loads; BS6399:Part 3:1988, Loadings for buildings: Code of Practice for imposed roof loads; CP3:Chapter V:Part 2:1972, Basic data for the design of buildings—wind loads.

The UK NAD has considered the effect of using the British Loading Codes in conjunction with EC2. It makes the following recommendations.

(a) The wind loads derived from CP3:Chapter V:Part 2 should be multiplied by 0·9.
(b) In calculating the vertical load on columns and walls, the reduction factors given in BS6399:Part 1:1984 may be used to allow for the probability of all the floors not being fully loaded simultaneously. This is in addition to the application of ψ factors to imposed loads (say in the case of wind and imposed load combinations), and is justified, as the ψ factors account for only the independent variable actions (in the above case, wind and imposed loads), not being at their full design values at the same time.
(c) Local drifting of snow should be regarded as an accidental load when considered in conjunction with dead load only. In combination with other actions, it should be taken as a normal imposed load.

Chapter 2

EC2 permits simplified load arrangements for the design of continuous beams and slabs without cantilevers that are subject to (a) predominantly uniformly distributed loads and (b) one variable action. The arrangements to be considered are

Clause 2.5.1.2.4

(a) alternate spans loaded with the design variable and permanent loads $(1 \cdot 35 G_k + 1 \cdot 5 Q_k)$ and other spans carrying only the design permanent load $(1 \cdot 35 G_k)$
(b) any two adjacent spans carrying the design variable and permanent loads $(1 \cdot 35 G_k + 1 \cdot 5 Q_k)$, with all other spans carrying only the design permanent load $(1 \cdot 35 G_k)$.

Although this is not stated, the above arrangements are intended for braced non-sway structures. They may also be used in the case of sway structures, but the following additional load cases involving the total frame will need to be considered.

(a) All spans loaded with the design permanent loads $(1 \cdot 35 G_k)$ and the frame subjected to the design wind load $(1 \cdot 5 W_k)$, where W_k is the characteristic wind load.
(b) All spans at all floor levels loaded with $(1 \cdot 35 G_k + 1 \cdot 35 Q_k)$ and the frame subjected to the design wind load of $1 \cdot 35 W_k$.
(c) In sensitive structures (i.e. sensitive to lateral deformation), it may be necessary to consider the effects of wind loading in conjunction with patterned imposed loading throughout the frame.

The simplified arrangements in section 3.2.3 above may also be applied to beam and slabs with cantilevers if the ratio length of the cantilever to the length of the

Table 3.1. Cantilever length for use of simplified load arrangement

Q_k/G_k	0·50	0·75	1·00	1·25	1·50	1·75	2·00	2·25
a/l	0·31	0·29	0·27	0·25	0·24	0·23	0·22	0·21

adjacent anchor span does not exceed the values given in Table 3.1, where a is the length of the cantilever and l is the length of the adjacent anchor span.

The reason for the limitation in section 3.2.3 above with regard to cantilevers is the effect on the position of the point of contraflexure.

It would therefore be acceptable to use the simplified arrangement of loads, provided the cantilever loads do not shift the point of the contraflexure further towards midspan than the location implied in a continuous beam subject to 'adjacent spans loaded' case. Table 3.1 has been derived on this basis. The point of contraflexure was calculated for (a) a continuous beam of four equal spans, loaded as shown in Fig. 3.2 and (b) a beam with a cantilever subject to the loading shown in Fig. 3.3. If the location of the point of contraflexure is to remain the same, it will then be possible to derive an expression for (a/l) as a function of the ratio of permanent and variable loads only.

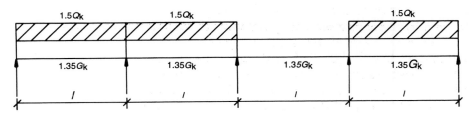

Fig. 3.2. Determination of point of contraflexure: continuous beam

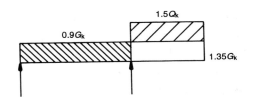

Fig. 3.3. Determination of point of contraflexure: beam with cantilever span

Clause 2.5.1.2(5) states that in linear elements and slabs subject predominantly to bending, the effects of shear and axial forces on deformation may be neglected if these are likely to be less than 10%. In practice, the designer need not actually calculate these additional deformations to carry out this check.

Deflections are generally of concern only in members with reasonably long spans. In such members, the contribution of shear to the deflections is never significant for members with normal (span/depth) ratios. When the spans are short, EC2 provides alternative design models (e.g. truss or strut-and-tie) in which deflections are rarely, if ever, a consideration.

The contribution of axial loads to deflections may be neglected if the axial stresses do not exceed $0·08f_{ck}$.

Clause 2.5.1.2(5)

Clause 4.3.1.2(6)

3.3. Imperfections

3.3.1. General
Perfection in buildings exists only in theory. In practice some degree of imperfection

is unavoidable, but designs should recognize this and ensure that buildings are sufficiently robust to withstand the consequences. For example, load-bearing elements may be out of plumb or the dimensional inaccuracies may cause eccentric application of loads. Most codes allow for these by prescribing a notional check for lateral stability. The exact approach adopted to achieve this differs between codes. In the UK, the structures are required to resist a notional horizontal load of at least 1·5% of the dead loads above any level being considered. If the factored wind loading exceeds the notional force the structures are designed for the wind loading rather than the notional force.

ENV EC2 has a number of provisions in this regard, affecting the design of (a) the structure as a whole, (b) some slender elements and (c) elements that transfer forces to bracing members.

3.3.2. Global analysis

For the analysis of the structure as a whole, an arbitrary inclination ν of the structure is prescribed, $\nu = 1/(100\sqrt{l})$ rad, but is not less than $1/400$ where the second-order effects are not significant, and not less than $1/200$ where they are. (In the UK, the NAD prescribes $1/200$ for both cases (see NAD Table 3).) In the above expression l is the total height of the structure in metres.

Clause 2.5.1.3(4)

In structures with a number of vertically continuous members, the degree of imperfection is statistically unlikely to be the same in all the members. This is recognized by a reduction factor $\alpha_n = \sqrt{[(1 + 1/n)/2]}$, where n is the number of vertically continuous elements (see Figs 3.4 and 3.5).

As a result of the inclination, a horizontal component of the vertical loads could be thought of being applied at each floor level, as shown in Figs 3.4 and 3.5. These horizontal forces should be taken into account in the stability calculation, unless they are smaller than design horizontal actions, in which case they can be ignored.

Clause 2.5.1.3(8)

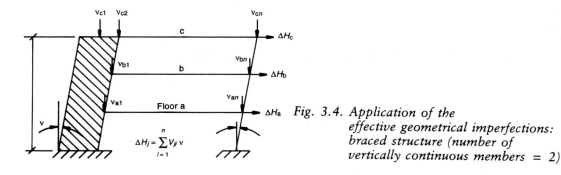

Fig. 3.4. Application of the effective geometrical imperfections: braced structure (number of vertically continuous members = 2)

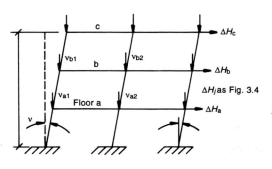

Fig. 3.5. Application of the effective geometrical imperfections: unbraced structure (number of vertically continuous members = 3)

3.3.3. Design of slender elements
In the design of slender elements that are prone to failure by buckling, e.g. slender columns, ENV EC2 requires geometrical imperfection to be added to other eccentricities. For example, in the design of the columns, an eccentricity of $vl_0/2$ is assumed for geometrical imperfection (where l_0 is the effective length of the column).

Clause 4.3.5.4

3.3.4. Members transferring forces to bracing elements
In the design of these elements (such as in a floor diagram), a force to account for the possible imperfection should be taken into account in addition to other design actions. This additional force is illustrated in Fig. 3.6. This force need not be taken into account in the design of the bracing element itself.

3.4. Second-order effects
As a structure subject to lateral loads deflects, the vertical loads acting on the structure produce additional forces and moments. These are normally referred to as second-order effects. Consider the cantilever column shown in Fig. 3.7. The deflection caused by the horizontal load alone is Δ_1. In this deflected state, the vertical load P will contribute a further bending moment and increase the lateral sway, and the final deflection will be Δ_2. This phenomenon is also commonly referred to as '$P\Delta$ effect'.

ENV EC2 requires second-order effects to be considered where they may significantly affect the stability of the structure as a whole or the attainment of ultimate limit state at critical sections.

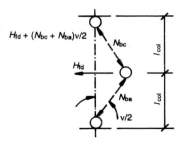

Fig. 3.6. Minimum tie force for perimeter columns

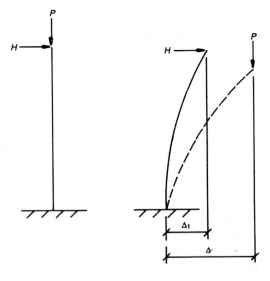

Fig. 3.7. Second-order effect

In the application rules, EC2 further states that, for normal buildings, second-order effects may be neglected if the bending moments caused by them do not increase the first-order bending moments (i.e. bending moments calculated ignoring the effects of displacements) by more than 10%.

Although this suggests that the designer will first have to check the second-order bending moment before ignoring it, EC2 provides some tests to verify whether or not second-order analysis is required. These tests are summarized in *Clause 4.3.5* and *Appendix 3* of EC2, and essentially ensure that adequate stiffness is provided. Only if the structure fails to satisfy the tests in EC2 is there a need to consult specialist literature (see e.g. Refs 1 and 2).

Clause 4.3.5
Appendix 3

In most practical building structures second-order effects are unlikely to be significant, as serviceability criteria to limit the lateral deflections will ensure that structures are not prone to $P\Delta$ effects.

3.5. Time-dependent effects

The main effects to be considered are creep and shrinkage of concrete and relaxation of prestressing steel. *Clause 4.4.1* provides information on when such effects need not be considered explicitly.

Clause 4.4.1

3.6. Design by testing

EC2 does not provide any usable guidance on this subject. EC1 is likely to contain some general guidance in the section 'Basis of design'.

Design entirely based on testing is not common in building structures. However, this is an accepted method in some other fields, e.g. pipes. In such work, test programmes should be designed in such a way that an appropriate design strength can be established, which includes proper allowance for the uncertainties covered by partial safety factors in conventional design. It will generally be necessary to establish the influence of material strengths on behaviour and their variability, so that a characteristic (and thus design) response can be derived. When testing is carried out on elements smaller than the prototype, size effects should be considered in the interpretation of results (e.g. for shear strength).

Testing may also be undertaken for other reasons, e.g.

(*a*) to appraise an existing structure
(*b*) to establish data for use in design
(*c*) to verify consistency of manufacture or performance of components.

Test methods and procedures will of course be different in each case. Before a test is planned, the precise nature of the information required from the test, together with criteria for judging the test, should be specified.

3.7. Structural analysis

3.7.1. Elastic analysis with or without redistribution

3.7.1.1. GENERAL.   ENV EC2 provides limited guidance on analysis. Elastic analysis remains the most popular method for frames (e.g. for moment distribution, slope deflection).

Braced frames may be analysed as a whole frame or may be partitioned into subframes. The subframes may consist of beams at one level with monolithic attachment to the columns. The remote ends may be assumed to be 'fixed', unless

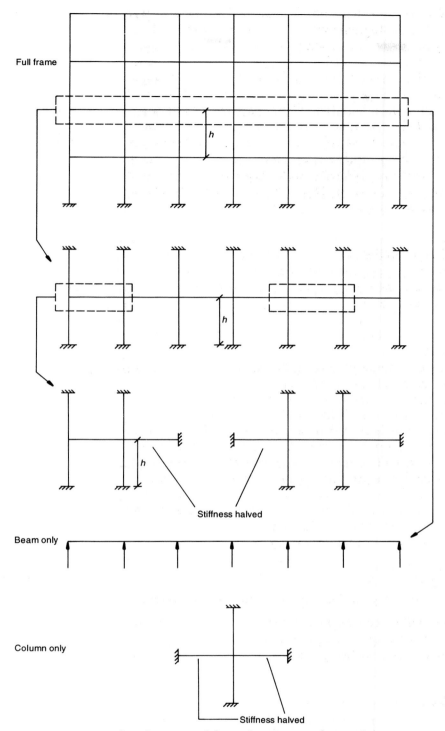

Fig. 3.8. Partitioning of multi-storeyed braced structures for analysis

a 'pinned' end is more reasonable in particular cases. As a further simplification, beams alone can be considered to be continuous over supports providing no restraint to rotation. Clearly this is more conservative (see Fig. 3.8).

In unbraced structures it is generally necessary to consider the whole structure, particularly when lateral loads are involved. A simplified analysis may be carried out assuming points of contraflexure at the mid-lengths of beams and columns (see Fig. 3.9). However, it should be remembered that this method will be inaccurate if (*a*) the feet of the column are not fixed, and/or (*b*) the beams and columns are not of similar stiffnesses.

3.7.1.2. STIFFNESS PARAMETERS. In the calculation of the stiffness of members it is normally satisfactory to assume a 'mean' value of modulus of elasticity for concrete (see *section 4.1.6.2* of the manual) and a moment of inertia based on uncracked gross cross-section of the member.

Section 4.1.6.2

3.7.1.3. EFFECTIVE SPANS. Calculations are performed using the effective spans defined below. The principle is to identify approximately the location of the line of reaction of the support

$$l_{\text{eff}} = l_n + a_1 + a_2$$

Clause 2.5.2.2.2

where l_{eff} is the effective span, l_n is the distance between the faces of supports, and a_1 and a_2 are the distances from the faces of the supports to the line of the effective reaction at the two ends of the member.

Typical conditions are considered below.

3.7.1.3.1. Case 1: Intermediate support over which member is continuous. a_j is the distance from the face of the support to its centre line (Fig. 3.10).

3.7.1.3.2. Case 2: Monolithic end support. a_j is the lesser of half the width of the support and half the overall depth of the member (Fig. 3.11).

3.7.1.3.3. Case 3: Discontinuous end support. a_j is the lesser of half the width of the support and one-third of the overall depth of the member (Fig. 3.12).

3.7.1.3.4. Case 4: Discontinuous end support on bearings. a_j is the distance from the face of the support to the centre line of the bearing (Fig. 3.13).

3.7.1.3.5. Case 5: Isolated cantilever. $a_j = 0$, i.e. the effective span is the length of the cantilever from the face of the support.

3.7.1.4. EFFECTIVE WIDTH OF FLANGES. As a result of 'shear lag', the stresses in the parts of a wide flange away from the web would be much lower than those at a flange/web junction. The calculation of the variation is a complex mathematical problem, therefore codes of practice allow approximations by which an 'effective' width can be calculated. A uniform distribution of stress is assumed over the effective width.

Clause 2.5.2.2.1

For building structures the effective widths shown in Fig. 3.14 can be used, where l_0 is the distance between points of contraflexure and is obtained from Fig. 3.15.

Note that (*a*) the length of the cantilever should be less than half the adjacent span; (*b*) the ratio of adjacent spans (other than the cantilever) should be between 1 and 1·5.

In analysis, EC2 permits the use of a constant flange width throughout the span. This should take the value applicable to span sections. This applies only to elastic design with or without redistribution; it does not apply when more rigorous non-linear methods are adopted.

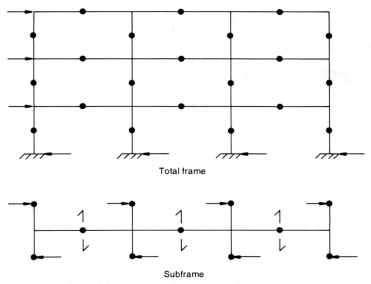

Fig. 3.9. *Simplified model for the analysis of unbraced structures*

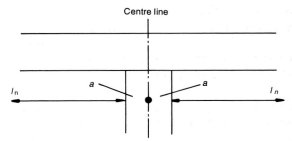

Fig. 3.10. *Intermediate support over which member is continuous*

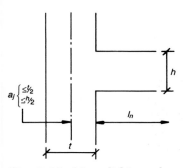 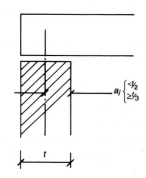

Fig. 3.11. *Monolithic end support*   Fig. 3.12. *Discontinuous end support*

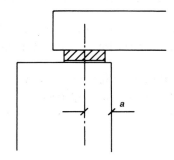

Fig. 3.13. *Discontinuous end support on bearings*

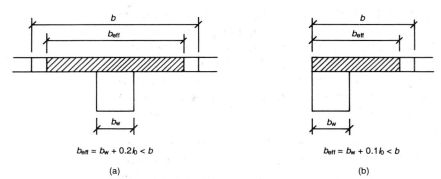

Fig. 3.14. *Effective flange width: (a) T-beam; (b) L-beam*

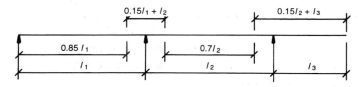

Fig. 3.15. *Definition of l_0 for the calculation of effective flange width*

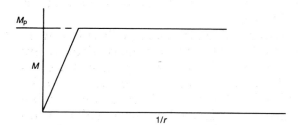

Fig. 3.16. *Moment/curvature: idealization*

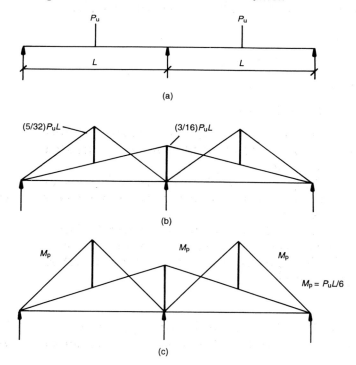

Fig. 3.17. *Bending moment at failure of (a) continuous beam, (b) an elastic beam, (c) an elasto-plastic beam*

3.7.1.5. REDISTRIBUTION OF MOMENTS.
The moment/curvature response of a true elasto-plastic material will typically be shown at Fig. 3.16.

The long plateau after M_p is reached implies a large rotation capacity. Consider a continuous beam made of such a material and loaded as shown in Fig. 3.17(a). When a bending moment in a critical section (usually at a support) reaches M_p, a plastic hinge is said to be formed. The structure will be able to withstand further increase in loading until sufficient plastic hinges form to turn the structure into a mechanism.

The bending moments at 'failure' will be as shown in Fig. 3.17(c). If the beam is elastic, the bending moment will be as shown in Fig. 3.17(b). A comparison of Fig. 3.17(c) and (b) shows that the elastic bending moment at the support has been reduced from $(3/16)P_u.L$ to $(1/6)P_u.L$, or redistributed by $\sim 11\%$.

The process illustrated in Fig. 3.17 is plastic analysis. Clearly, for plasticity to be fully exploited, the material must possess adequate ductility (rotation capacity). Concrete has only limited capacity in this regard. Moment redistribution procedure is an allowance for the plastic behaviour without carrying out plastic hinge analysis. Indirectly, it also ensures that the yield of sections under service loads and large uncontrolled deflections are avoided.

ENV EC2 allows moment redistribution of up to 30% in braced structures. No redistribution is permitted in sway frames or in situations where rotation capacity cannot be defined with confidence.

When redistribution is carried out, it is essential to maintain equilibrium between the applied loads and the resulting distribution of bending moments. Thus, where support moments are reduced, the moments in the adjacent spans will need to be increased to maintain equilibrium for that particular arrangement of loads.

In EC2 the limit to the amount of redistribution is related to the ductility characteristics (see *section 4.2.3* of the manual) of the reinforcement used: for high-ductility steel, 30%; for normal-ductility steel, 15%. EC2 permits redistribution to be carried out without the rotation being explicitly checked, provided the following expressions are satisfied.

Clause 2.5.3.4.2
Section 4.2.3

For concrete grade $\not> C35/45$

$$\delta \geq 0.44 + 1.25\, x/d$$

and for concrete grade $> C35/45$

$$\delta \geq 0.56 + 1.25\, x/d$$

where δ is the ratio of the redistributed moment to the moment before redistribution, x is the depth of the neutral axis at the ultimate limit state after redistribution, and d is the effective depth.

These requirements are shown in Fig. 3.18. Table 3.2 gives the x/d values for typical values of δ. *Section 5.2* of the manual gives design equations and charts that take account of the above limitations.

Section 5.2

Note that the above limitations are an attempt to ensure sufficiently ductile behaviour. Higher strength concrete tends to be brittle; hence the more onerous limits on x/d. The ratio x/d also increases with steel content and therefore indirectly affects the ductility of the section.

3.7.2. Plastic analysis
Apart from moment redistribution, ENV EC2 allows the use of plastic analysis without any direct check on rotation capacity provided the following conditions are satisfied.

Clause 2.5.3.5.5
Appendix 2.4(2)

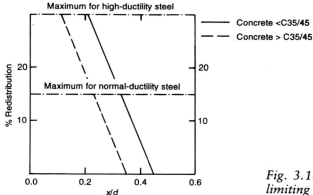

Fig. 3.18. Redistribution of moment and limiting x/d values

Table 3.2. Limiting neutral axis depths and redistribution of moment

δ	x/d For concrete grade ≯ C35/45	x/d For concrete grade > C35/45
0·70	0·208	0·112
0·80	0·288	0·192
0·85	0·328	0·232
0·90	0·368	0·272

(a) The area of reinforcement at any location in any direction should not exceed a value corresponding to $x/d = 0·25$. EC2 does not provide any basis for this value, which might be at odds with Table 3.2 above. It is believed that some countries have used this value in practice with satisfactory results. However, this experience is generally related to slabs, which are more ductile than other members.

(b) Normal-ductility steel should not be used unless specially justified.

(c) The ratio of the moments over continuous edges to the moments in the span should be between 1·0 and 2·0.

Condition (a) can be expressed as a reinforcement percentage for a balanced section, i.e. when both concrete and reinforcement have reached their limiting strains (Fig. 3.19).

$$F_c = F_s$$
$$F_c = (0·8x)(0·85 f_{ck}/1·5)b$$

Substituting $x = 0·25d$

$$F_c = 0·113 \, bd f_{ck}$$
$$F_s = A_s(0·87 f_y)$$

Substituting $A_s = (\rho/100)bd$ and equating F_c and F_s

$$\rho = 12·99 f_{ck}/f_y$$

Table 3.3 gives the values of ρ for various combinations of f_{ck} and f_y.

The most common plastic method used in practice is the yield line analysis for slabs. For details of this method, standard textbooks should be consulted.

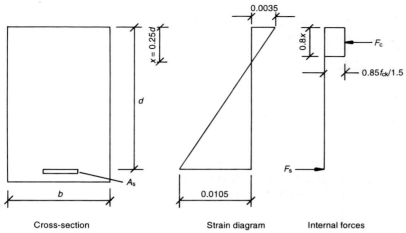

Fig. 3.19. Strain and stress conditions in a balanced section

Table 3.3. Values of $\rho = 100\,A_s/bd$ for plastic analysis

f_{ck}:N/mm²	f_y:N/mm²		
	500	460	250
20	0·52	0·56	1·04
25	0·65	0·71	1·30
30	0·78	0·85	1·56
35	0·91	0·99	1·82

3.7.3. Non-linear analysis

This takes into account the non-linear deformation properties of reinforced concrete sections. This method is seldom used in practical design, as it is complex and requires a computer and prior knowledge of reinforcement details throughout the structure. It could be useful in appraising the capacity of existing structures or when a large repetition of a particular structure is considered (e.g. precast frames).

The method involves plastic hinge analysis (see section 3.7.1.5) which includes explicit calculation of rotations at hinges. In this method, 'failure' is reached when (a) limiting rotation occurs at a plastic hinge or (b) sufficient hinges form to render the structure into a mechanism, whichever occurs first. A hinge is said to have formed when the steel starts to yield.

Consider the encastre beam shown in Fig. 3.20. The procedure to predict the ultimate load capacity may be as follows.

(a) Knowing the reinforcement, section properties and concrete grade, calculate the M_{yk} at A and C. M_{yk} is the bending moment that produces a stress f_{yk} in the tension reinforcement. Also calculate M_{yd}, by applying γ_m.

(b) Calculate the loading that will produce an elastic bending moment A and B equal to M_{yk}.

(c) Increase the loading in stages. At each stage calculate: (i) the rotation of the hinges A and B by integrating the curvatures of the beam between hinges (this will require the beam to be divided into a number of sections, at each of which the curvature should be calculated using *equations (A2.1) and (A2.2) in Appendix 2*); (ii) the bending moment at C; (iii) the strains in steel and concrete at A, B and C.

Appendix 2

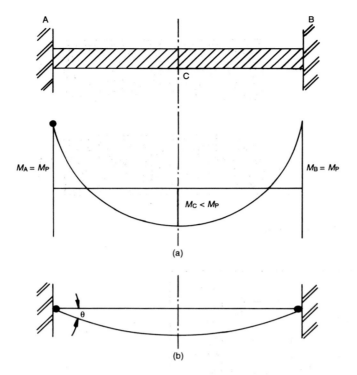

Fig. 3.20. Plastic analysis of an encastre beam: (a) elastic moments when the first hinge(s) form at A and B; (b) rotation θ at the first hinge when the load is increased

(d) Compare the rotation at A and B with the limiting rotation given in *Fig. A2.2* in *Appendix 2*.

Appendix 2

(e) Failure is reached when either the limiting rotation is reached at A and B or M_{yk} is reached at C.

(f) Calculate the load corresponding to 'failure' as noted above.

Note that the rotations are calculated using characteristic values of mechanical properties, whereas the strengths are calculated using design properties, i.e. γ_m is used.

The above procedure is set out to clarify the steps involved in a relatively simple example. The analysis of a continuous beam will obviously be fairly complex even with a computer. Also, a number of boundary conditions will need to be imposed in the analysis. In practice, therefore, this procedure is rarely used, and elastic analysis with moment redistribution is preferred.

3.7.4. Strut-and-tie models

Strut-and-tie models utilize the lower bound theorem of plasticity, which can be summarized as follows:

'For a structure under a given system of external loads, if a stress distribution throughout the structure can be found such that (i) all conditions of equilibrium are satisfied and (ii) the yield condition is not violated anywhere, then the structure is safe under the given system of external loads.'

This approach particularly simplifies the analysis of the parts of the structure where linear distribution of strain is not valid. Typical areas of application are noted in section 3.1 above.

Typical models are shown in Fig. 3.21. As can be seen, the structure is thought

ANALYSIS

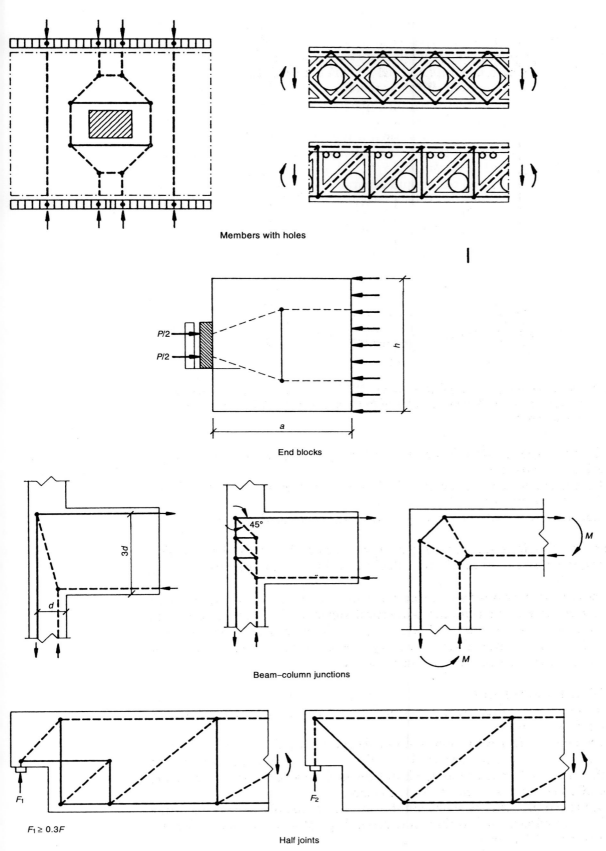

Fig. 3.21. Typical strut-and-tie model

of as comprising notional concrete struts and reinforcement ties. Occasionally concrete ties may also be considered (e.g. slabs without stirrups, anchorages without transverse reinforcement). While there may appear to be infinite freedom to choose the orientation of struts and ties, this is not in fact so. Concrete has only limited plastic deformation capacity, therefore the model has to be chosen with care to ensure that the deformation capacity is not exceeded at any point before the assumed states of stress are reached in the structure. Generally it will be safe to model the struts and ties in such a way that they closely follow the 'stress paths' indicated by the theory of elasticity. In EC2, some geometrical limitations are introduced (e.g. *clause 2.5.3.7.2*) to achieve the same end. The angles between the struts and ties should generally be greater than 45° in order to avoid incompatibility problems. Where several models are possible, the 'correct' model is the one with the least number of internal members and the least deformation. In this context the deformation of the struts can be neglected and the model optimized by minimizing the expression

Clause 2.5.3.7.2

$$\Sigma F_i l_i \epsilon_{mi}$$

where F_i is the force in the tie, l_i is the length of the tie and ϵ_{mi} is the mean strain in tie i.

Having idealized the structure as struts and ties, it is a simple matter to arrive at the forces in them based on equilibrium with external loads.

Limiting stresses are

(a) in the ties, f_{yd} (design value of yield stress)
(b) in struts subject to uniaxial stresses, $0 \cdot 6 f_{cd}$ (f_{cd} is the design value of concrete cylinder strength)
(c) when a triaxial state of stress can be achieved in struts, $1 \cdot 0 f_{cd}$
(d) under concentrated load $3 \cdot 3 \times (0 \cdot 85 f_{cd})$.
Note that the actual stress that should be compared with this limiting value is

Clause 5.4.8.1

$$0 \cdot 85 f_{cd} (A_{c1}/A_{c0})$$

where A_{c0} is the area over which the load is distributed uniformly and A_{c1} is the maximum concentric area that it is possible to inscribe in the total area of the member in the same plane as the loaded area.

In general, if the above limits are observed, the 'node' regions are unlikely to be critical. Schlaich and Schafer[3] recommend a procedure for checking the node regions, in their paper 'Design and detailing of structural concrete using strut-and-tie models'.

Bearing in mind that strut-and-tie models come under plastic analysis, it is interesting to note that EC2 does not impose any conditions similar to those noted in section 3.7.2 above. This is an inconsistency. It is therefore important to follow the theory of elasticity fairly closely in choosing the model, as discussed above.

3.8. Design aids and simplifications

3.8.1. One-way spanning slabs and continuous beams

The bending moment coefficients shown in Fig. 3.22(a) can be used for three or more equal spans subject to equal uniformly distributed loads. These coefficients assume supports that do not offer rotational restraint. The coefficients for locations 3–6 may be used (without sacrificing too much economy) even when continuity with internal columns is to be taken into account. However, continuity with

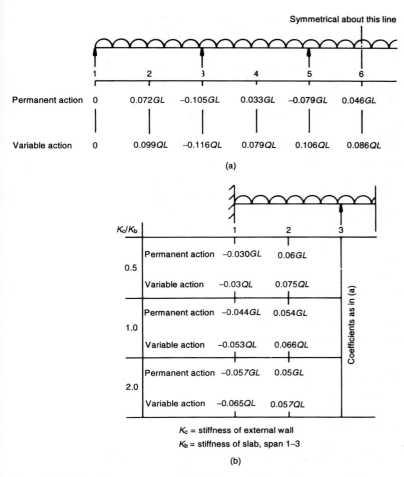

Fig. 3.22. Bending moment coefficients: (a) continuous slab on point supports: (b) continuous slabs with continuity at external columns

external columns has a significant effect on the bending moments at locations 1 and 2. The modifications shown in Fig. 3.22(b) may be made, depending on the ratio of the total column wall stiffness to the slab stiffness.

Note that the values of bending moments in Fig. 3.22 for variable loads are the maximum values obtained by a linear analysis considering the cases of loading alternate and adjacent spans. It will not therefore be possible to carry out a moment redistribution, as the loading case applicable will vary with each span.

A further approximation may be made if

(a) the characteristic variable load does not exceed the characteristic permanent load
(b) loads are predominantly uniformly distributed over three or more spans
(c) variations in span lengths do not exceed 15% of the longest span.

Under these conditions the bending moments and shear forces can be obtained by use of Table 3.4.

Table 3.4 assumes that the ratio of the stiffness of the end columns to that of the slab (or beam) does not exceed 1·0. No redistribution of moments is permitted to the bending moments obtained by use of Table 3.4.

Table 3.4. Ultimate bending moment and shear forces in one-way spanning slabs (continuous beams and flat slabs)

	End support and end span				At first interior support	Middle of interior spans	Interior supports
	Simple		Monolithic				
	At outer support	Near middle of end span	At outer support	Near middle of end span			
Moment	0	0.086 Fl	−0.040 Fl	0.075 Fl	−0.086 Fl	−0.063 Fl	−0.063 Fl
shear	0.40 F	—	0.46 F	—	0.6 F	—	0.50 F

F is the total design ultimate load ($1.35G_k + 1.50Q_k$); Fl is the effective span.

3.8.2. Two-way spanning slabs

3.8.2.1. GENERAL. For rectangular slabs with standards edge conditions and subject to uniformly distributed loads, the bending moments are normally obtained by use of tabulated coefficients. Such coefficients are provided below.

For slabs with irregular plan shapes and slabs subject to a combination of point loads and distributed loads, Johansen's yield line analysis and the Hillerborg strip method provide powerful methods for strength calculations.[4-6]

3.8.2.2. SIMPLY SUPPORTED SLABS. Where (a) corners of slabs are free to lift and (b) no provision is made to resist forces at the corners, the maximum moments per unit width are given by

$$M_{sx} = \text{bending moment in strips with span } l_x$$
$$= \alpha_{sx} q \, l_x^2$$
$$M_{sy} = \text{bending moment in strips with span } l_y$$
$$= \alpha_{sy} q \, l_x^2$$

where l_x is the shorter span of the panel, l_y is the longer span of the panel and q is the design ultimate load per unit area. Values of α_{sx} and α_{sy} are given in Table 3.5 for various ratios of l_y and l_x.

3.8.2.3. RECTANGULAR PANELS WITH RESTRAINED EDGES. Where corners are (a) prevented from lifting and (b) reinforced to resist torsion, the maximum bending moments per unit width are given by

$$M_{sx} = \beta_{sx} q \, l_x^2$$
$$M_{sy} = \beta_{sy} q \, l_x^2$$

Table 3.5. Bending moment coefficients for slabs spanning in two directions at right-angles, simply supported on four sides

l_y/l_x	1.0	1.1	1.2	1.3	1.4	1.5	1.75	2.0
α_{sx}	0.062	0.074	0.084	0.093	0.099	0.104	0.113	0.118
α_{sy}	0.062	0.061	0.059	0.055	0.051	0.046	0.037	0.029

where M_{sx} is the maximum design moment either over supports or at mid-span on strips with span l_x, M_{sy} is the maximum design moment either over supports or at mid-span on strips with span l_y, q is the design ultimate load per unit area, l_x is the shorter span and l_y is the longer span.

For various ratios (l_y/l_x), values of β_{sx} and β_{sy} are given in Table 3.6 for various edge conditions.

The following are the conditions for use of the Table 3.6 value.

(a) Table 3.6 is based on the yield line, i.e. plastic analysis. Therefore the conditions in section 3.7.2(a)–(c) above should be observed.
(b) The permanent and variable loads in adjacent panels should be approximately the same as on the panel being designed.
(c) The span of the adjacent panels in the direction perpendicular to the line of the common support should be approximately the same as the span of the panel being designed.
(d) Corners at the junction of simply supported edges should be reinforced as shown in Fig. 3.23.
(e) The panel should observe the detailing rules in Chapter 10 below.

In the corner area shown:

(a) provide top and bottom reinforcement
(b) in each layer provide bars parallel to the slab edges
(c) in each of the four layers the area of reinforcement should be equal to 75% of the reinforcement required for the maximum span moment
(d) the area of reinforcement in (c) can be halved if one edge of the slab in the corner is continuous.

3.8.2.4. UNEQUAL EDGE CONDITIONS IN ADJACENT PANELS.

In some cases, the bending moments at a common support, obtained by considering the two adjacent panels in isolation, may differ significantly (say by 10%), because of the differing edge conditions at the far supports or differing span lengths or loading.

Consider panels 1 and 2 in Fig. 3.24. As the support on grid A for panel 1 is discontinuous and the support on grid C for panel 2 is continuous, the moments for panels 1 and 2 for the support on grid B could be significantly different. In these circumstances, the slab can be reinforced throughout for the worst case span and support moments. However, this may be uneconomic in some cases. In such cases, the following distribution procedure can be used.

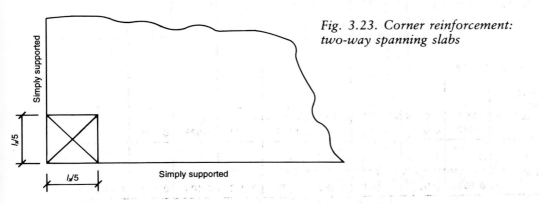

Fig. 3.23. Corner reinforcement: two-way spanning slabs

Table 3.6. Bending moment coefficients for rectangular panels supported on four sides with provision for torsion at corners

Type of panel and moments considered	Short span coefficients β_{sx} l_y/l_x								Long span coefficients β_{sy}, for all values of l_y/l_x
	1·0	1·1	1·2	1·3	1·4	1·5	1·75	2·0	
Interior panels Negative moment at continuous edge	0·031	0·037	0·042	0·046	0·050	0·053	0·059	0·063	0·032
Positive moment at mid-span	0·024	0·028	0·032	0·035	0·037	0·040	0·044	0·048	0·024
One short edge discontinuous Negative moment at continuous edge	0·039	0·044	0·048	0·052	0·055	0·058	0·063	0·067	0·037
Positive moment at mid-span	0·029	0·033	0·036	0·039	0·041	0·043	0·047	0·050	0·028
One long edge discontinuous Negative moment at continuous edge	0·039	0·049	0·056	0·062	0·068	0·073	0·082	0·089	0·037
Positive moment at mid-span	0·030	0·036	0·042	0·047	0·051	0·055	0·062	0·067	0·028
Two adjacent edges discontinuous Negative moment at continuous edge	0·047	0·056	0·063	0·069	0·074	0·078	0·087	0·093	0·045
Positive moment at mid-span	0·036	0·042	0·047	0·051	0·055	0·059	0·065	0·070	0·034
Two short edges discontinuous Negative moment at continuous edge	0·046	0·050	0·054	0·057	0·060	0·062	0·067	0·070	—
Positive moment at mid-span	0·034	0·038	0·040	0·043	0·045	0·047	0·050	0·053	0·034
Two long edges discontinuous Negative moment at continuous edge	—	—	—	—	—	—	—	—	0·045
Positive moment at mid-span	0·034	0·046	0·056	0·065	0·072	0·078	0·091	0·100	0·034
Three edges discontinuous (one long edge continuous) Negative moment at continuous edge	0·057	0·065	0·071	0·076	0·081	0·084	0·092	0·098	—
Positive moment at mid-span	0·043	0·048	0·053	0·057	0·060	0·063	0·069	0·074	0·044
Three edges discontinuous (one short edge continuous) Negative moment at continuous edge	—	—	—	—	—	—	—	—	0·058
Positive moment at mid-span	0·042	0·054	0·063	0·071	0·078	0·084	0·096	0·105	0·044
Four edges discontinuous Positive moment at mid-span	0·055	0·065	0·074	0·081	0·087	0·092	0·103	0·111	0·056

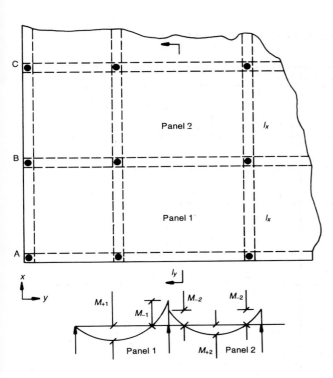

Fig. 3.24. Two-way spanning slabs: unequal edge condition in adjacent panels

(a) Obtain the support moments for panels 1 and 2 from Table 3.6. If M_{-1} and M_{-2} are treated as fixed end moments, the moments may be distributed in proportion to the stiffnesses of span l_x in panels 1 and 2. Thus a revised bending moment M'_{-B} may be obtained for support over B.

(b) The span moments in panels 1 and 2 should be recalculated as follows

$$M'_{+1} = (M_{-1} + M_{+1}) - M'_{-B}$$
$$M'_{+2} = (M_{-2} + M_{-2} + M_{+2}) - M'_{-B} - M_{-2}$$

(Note that this assumes that the final moment over C is M_{-2})

(c) For curtailment of reinforcement, the point of contraflexure can be obtained by assuming a parabolic distribution of moments in each panel.

3.8.2.5. LOADS ON SUPPORTING BEAMS.

Loads on supporting beams can be obtained by use of either Fig. 3.25 or Table 3.7.

Note that (a) the reactions shown apply when all edges are continuous (or

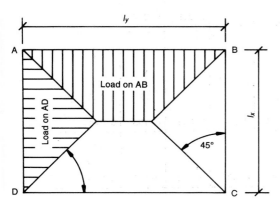

Fig. 3.25. Loads on beams supporting two-way spanning slabs

discontinuous); (*b*) when one edge is discontinuous, the reactions on all continuous edges should be increased by 10% and the reaction on the discontinuous edge can be reduced by 20%; (*c*) when adjacent edges are discontinuous, the reactions should be adjusted for elastic shear considering each span separately.

3.8.3. Flat slabs: methods of analysis
Flat slab structures are defined as slabs (solid or coffered) supported on point supports. Unlike two-way spanning slabs on line supports, flat slabs can fail by

Table 3.7. Shear force coefficients for uniformly loaded rectangular panels supported on four sides with provision for torsion at corners

Type of panel and location	β_{vx} for values of γ_y/γ_x								β_{vy}
	1·0	1·1	1·2	1·3	1·4	1·5	1·75	2·0	
Four edges continuous Continuous edge	0·33	0·36	0·39	0·41	0·43	0·45	0·48	0·50	0·33
One short edge discontinuous Continuous edge Discontinuous edge	0·36 —	0·39 —	0·42 —	0·44 —	0·45 —	0·47 —	0·50 —	0·52 —	0·36 0·24
One long edge discontinuous Continuous edge Discontinuous edge	0·36 0·24	0·40 0·27	0·44 0·29	0·47 0·31	0·49 0·32	0·51 0·34	0·55 0·36	0·59 0·38	0·36 —
Two adjacent edges discontinuous Continuous edge Discontinuous edge	0·40 0·26	0·44 0·29	0·47 0·31	0·50 0·33	0·52 0·34	0·54 0·35	0·57 0·38	0·60 0·40	0·40 0·26
Two short edges discontinuous Continuous edge Discontinuous edge	0·40 —	0·43 —	0·45 —	0·47 —	0·48 —	0·49 —	0·52 —	0·54 —	— 0·26
Two long edges discontinuous Continuous edge Discontinuous edge	— 0·26	— 0·30	— 0·33	— 0·36	— 0·38	— 0·40	— 0·44	— 0·47	0·40 —
Three edges discontinuous (one long edge continuous) Continuous edge Discontinuous edge	0·45 0·30	0·48 0·32	0·51 0·34	0·53 0·35	0·55 0·36	0·57 0·37	0·60 0·39	0·63 0·41	— 0·29
Three edges discontinuous (one short edge continuous) Continuous edge Discontinuous edge	— 0·29	— 0·33	— 0·36	— 0·38	— 0·40	— 0·42	— 0·45	— 0·48	0·45 0·30
Four edges discontinuous Discontinuous edge	0·33	0·36	0·39	0·41	0·43	0·45	0·48	0·50	0·33

See document EC2/TLLCH3.TAB for tables 3.7 and 3.6

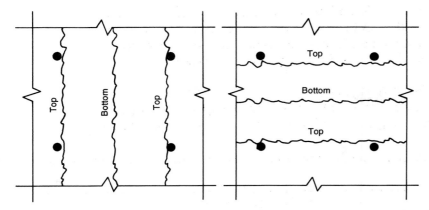

Fig. 3.26. Possible failure modes of flat slabs

yield lines in either of the two orthogonal directions (see Fig. 3.26). For this reason, flat slabs must be capable of carrying the total load on the panel in *each* direction.

EC2 provides virtually no information for the analysis of flat slabs. The recommendations noted here follow closely the provisions in the UK Code of Practice BS8110, which have given satisfactory results in practice over a long period.

EC2 limits itself to punching shear around columns. However, many recognized methods are available, including

(*a*) equivalent frame method
(*b*) simplified coefficients
(*c*) yield line analysis
(*d*) grillage analysis.

In the manual, only methods (*a*) and (*b*) are considered. For other approaches, specialist literature should be consulted, e.g. CIRIA Report 110.[7]

3.8.3.1. EQUIVALENT FRAME METHOD.
3.8.3.1.1. Division into frames. The structure should be divided in two orthogonal directions into frames consisting of columns and strips of slabs acting as 'beams'. The width of the slab to be used for assessing the stiffness depends on the aspect ratio of the panels and whether the loading is vertical or horizontal.

(*a*) *Vertical loading:* when the aspect ratio is less than 2, the width may be taken as the distance between the centre lines of the adjacent panels. For aspect ratios greater than 2, the width may be taken as the distance between the centre lines of the adjacent panels when bending is considered in the direction of longer length spans of the panel, and twice this value for bending in the perpendicular direction (see Fig. 3.27).

Width of beams for frame analysis:

When $l_y < 2l_x$

$$W_x = (l_{x1} + l_{x2})/2$$
$$W_y = (l_{y1} + l_{y2})/2$$

When $l_y > 2l_x$

$$W_x = (l_{x1} + l_{x2})/2$$
$$W_y = 2W_x$$

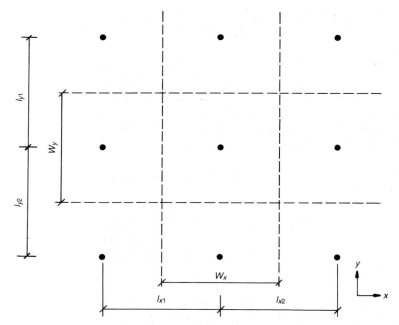

Fig. 3.27. Flat slabs: division into frames

(b) *Horizontal loading:* horizontal loading in the frame will be considered only in unbraced structures. In these cases, the question of restraint to the columns, and hence the effective length of columns, is a matter of judgement. If the stiffness of the slab framing into the column is overestimated, the effective length of the column will reduce correspondingly. As the stiffness at the slab–column junction is a grey area, codes of practice adopt a cautious approach. For the slab, half the stiffness applicable to vertical loading is used.

Stiffness properties are generally based on the gross cross-section (ignoring the reinforcement). Additional stiffening effects of drops or solid concrete around columns in coffered slabs may be included, but this will complicate hand calculations. Drops (and solid areas) should be taken into account only when the smaller dimension of the drop (and solid areas) is at least 33% of the smaller dimension of the surrounding panels.

3.8.3.1.2. *Analysis.* The equivalent frames can be analysed using any of the standard linear elastic methods, e.g. moment distribution (see section 3.7.1). Braced structures may be partitioned into subframes consisting of the slab at one level continuous with columns above and below. The far ends of the columns are normally taken as fixed unless this assumption is obviously wrong, for example in columns with small pad footings not designed to take moments.

Load combinations given in section 3.2.3 can be used.

The bending moments obtained from the analysis should be distributed laterally in the 'width' of the slab in accordance with section 3.8.3.2.3 below.

3.8.3.2. SIMPLIFIED COEFFICIENTS. In braced buildings with at least three approximately equal bays and both slabs subject predominantly to uniformly distributed loads, the bending moments and shear forces can be obtained using the coefficients given in Table 3.4.

The bending moments obtained from Table 3.4 should be distributed laterally in the 'width' of the slab in accordance with section 3.8.3.2.3 below.

3.8.3.2.1. *Lateral distribution of moments in the width of the slab.* In order to control the cracking of the slabs in service conditions, the bending moments obtained from the analysis should be distributed taking into account the elastic behaviour of the slab. Not surprisingly, the strips of the slab on the lines of the columns will be stiffer than those away from the columns. Thus the strips closer to the column lines will attract higher bending moments.

3.8.3.2.2. *Division of panels.* Flat slab panels should be divided into column and middle strips as shown in Fig. 3.28.

As drawn, Fig. 3.28 applies to slabs without drops (or solid areas around columns in coffered slabs). When drops (or solid areas around columns in coffered slabs) of plan dimensions greater than $l_x/3$ are used, the width of the column strip can be taken as the width of the drop. The width of the middle strip should be adjusted accordingly.

3.8.3.2.3. *Allocation of moments between strips.* The bending moments obtained from the analysis should be distributed between the column and middle strips in the proportions given in Table 3.8.

In some instances the analysis may show that hogging moments occur in the centre of a span (e.g. the middle span of a three-bay structure, particularly when it is shorter than the adjacent spans). The hogging moment can be assumed to be uniformly distributed across the slab if the negative moment at mid-span is less than 20% of the negative moment at supports. When this condition is not met, the moment is concentrated more in the middle strip.

3.8.3.2.4. *Moment transfer at edge columns.* As a result of flexural and torsional

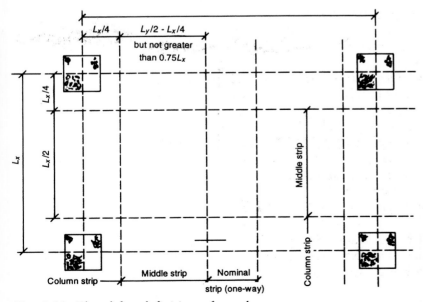

Fig. 3.28. Flat slabs: definition of panels

Table 3.8. Distribution of design moments in panels of flat slabs

	Apportionment between column and middle strip expressed as percentages of the total negative or positive design moment	
	Column strip	Middle strip
Negative	75	25
Positive	55	45

For the case where the width of the column strip is taken as equal to that of the drop, and the middle strip is thereby increased in width, the design moments to be resisted by the middle strip should be increased in proportion to its increased width. The design moments to be resisted by the column strip can be decreased by an amount such that the total positive and the total negative design moments resisted by the column strip and middle strip together are unchanged.

cracking of the edge (and corner) columns, the effective width through which moments can be transferred between the slabs and the columns will be much narrower than in the case of internal columns. Empirically, this is allowed for in design by limiting the maximum moment that the slab (without edge beams) can transfer to the columns.

$$M_{tmax} = 0.167\, b_e d^2 f_{ck}$$

for concrete grades C35/45 or less

$$M_{tmax} = 0.136\, b_e d^2 f_{ck}$$

for concrete grades C40/50 or greater.

where b_e is the effective width of the strip transferring the moment as defined in Fig. 3.29, and d is the effective depth of the slab.

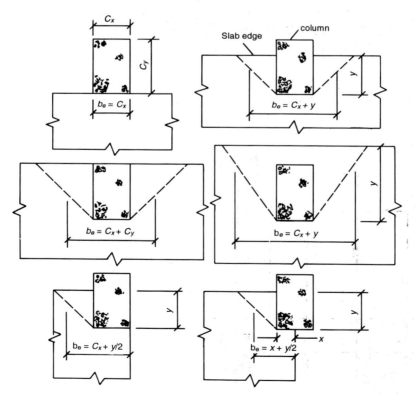

Fig. 3.29. *Effective slab widths for moment transfer*

M_{tmax} should not be less than 50% of the design moment obtained from an elastic analysis or 70% of the design moment obtained from a grillage or finite element analysis. If M_{tmax} is less than these limits the structure should be redesigned.

When the bending moment at the other support obtained from the analysis exceeds M_{tmax}, the moment at the outer support should be reduced to M_{tmax} and the span moment should be increased accordingly.

The reinforcement required in the slab to transfer the outer support moment to the column should be placed on a width of slab $(C_x + 2r)$ (see Fig. 3.30). For slabs with thickness less than 300 mm, $r = C_y$; for thicker slabs, $r = 1 \cdot 67 C_y$. In the latter case it is essential to provide torsional links along the edge of the slab. However, U-bars (as distinct from L-bars) with longitudinal anchor bars in the top and bottom can be assumed to provide the necessary torsional reinforcement (see Fig. 3.31).

Bending moments in excess of M_{tmax} may be transferred to the column only if an edge beam (which may be a strip of slab) is suitably designed to resist the torsion.

3.8.4. Beams

For the design of beams, the bending moment coefficients given in Fig. 3.22 and Table 3.4 may be used, noting the conditions to be complied with, which are discussed in section 3.8.1.

3.8.5. Simplifications

ENV EC2 permits the following simplifications regardless of the method of analysis used.

(a) A beam or a slab that is continuous over a support assumed to offer no restraint to rotation (e.g. over walls) may be designed for a support moment less than the moment theoretically calculated on the centre line of the support. The permitted reduction in moment is equal to $F_s b_{supp}/8$,

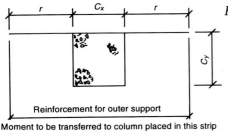

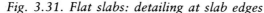

Fig. 3.30. Flat slabs: detailing at outer support

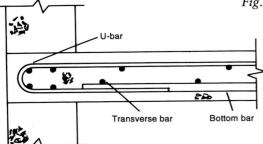

Fig. 3.31. Flat slabs: detailing at slab edges

where F_s is the design support reaction and b_{supp} is the breadth of the support. This recognizes the effect of the width of support and arbitrarily rounds off the peak in the bending moment diagram.

(b) Where a beam or a slab is cast monolithically into its supports, the critical moment may be taken as that at the face of the supports (but see also (c) below). This provision is reasonable, as flexural failure cannot occur within the support.

(c) The design moment at the faces of rigid supports should not be less than 65% of the support moment calculated assuming full fixity at the faces of support. This ensures a minimum design value for the support moment, particularly in the case of wide supports.

(d) Loads on members supporting one-way spanning continuous slabs (solid and ribbed) and beams (including T-beams) may be assessed on the assumption of simple supports for the supported elements, except that the effect of continuity on reactions should be allowed for (i) at the first internal support and (ii) at any other internal support where the spans on either side differ by more than 30%. This is a pragmatic rule and will not lead to significant inaccuracies.

CHAPTER 4

Materials and design data

4.1. Concrete

4.1.1. General
Basic information on material properties is given in *Chapters 3 and 4* of the Code. EC2 relies on ENV 206 for the specification of technological aspects of concrete, including tests for confirming the properties.

Chapter 3
Chapter 4

4.1.2. Strength classes and compressive strength f_{ck}
In EC2 the compressive strength of concrete is classified by concrete strength classes, which relate to the cylinder strength f_{ck} (or the equivalent cube strength). Thus, strength class C30/37 denotes concrete with cylinder strength of 30 N/mm² and cube strength of 37 N/mm². Fig. 4.1 shows the relationship of the cylinder and cube strengths.

EC2 is based on the characteristic compressive strength f_{ck}, defined as the value below which 5% of all possible strength test results for the specified concrete may be expected to fall.

Special justification of properties is required if strength classes weaker than C12 and stronger than C50 are to be used for reinforced or prestressed concrete work. Concrete grades weaker than C12 are very rare, except perhaps in mass concrete applications. However, there has been considerable research into high-strength concrete, particularly in the USA. Therefore justification of properties for concrete grades up to about C80 should not be a great problem.

4.1.3. Tensile strength f_{ct}
Tensile strength is defined in EC2 as the maximum stress that concrete can withstand when subjected to uniaxial tension. When the tensile strength is determined as splitting tensile strength or flexural tensile strength, the following relationships apply

$$f_{ct} = 0 \cdot 9 f_{ct}, \text{ splitting}$$

$$f_{ct} = 0 \cdot 5 f_{ct}, \text{ flexural}$$

In the absence of more accurate data, the values for tensile strength given in Table 4.1 may be used.

4.1.4. Shear strength τ_{Rd}
The basic shear strength of concrete may be obtained from Table 4.2 for various values of f_{ck}. Table 4.2 was derived using the relationship

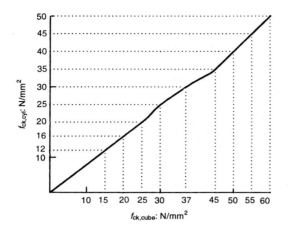

Fig. 4.1. Relation of cylinder strength and cube strength

Table 4.1 Tensile strength of concrete

Strength class of	C12/15	C16/20	C20/25	C25/30	C30/37	C35/45	C40/50	C45/55	C50/60
f_{ck}	12	16	20	25	30	35	40	45	50
f_{ctm}	1·6	1·9	2·2	2·6	2·9	3·2	3·5	3·8	4·1
$f_{ctk}0\cdot05$	1·1	1·3	1·5	1·8	2·0	2·2	2·5	2·7	2·9
$f_{ctk}0\cdot95$	2·0	2·5	2·9	3·3	3·8	4·2	4·6	4·9	5·3

Notes

f_{ctm} is the mean value of the tensile strength $= 0\cdot3\,(f_{ck})^{0\cdot67}$

$f_{ctk0\cdot06}$ is the lower characteristic tensile strength (5% fractile) $= 0\cdot7 f_{ctm}$

$f_{ctk0\cdot95}$ is the upper characteristic tensile strength (95% fractile) $= 1\cdot3 f_{ctm}$

Table 4.2. Values of τ_{Rd} with $\gamma = 1\cdot5$

f_{ck}: N/mm²	12	16	20	25	30	35	40	45	50
τ_{Rd}: kN/mm²	0·18	0·22	0·26	0·30	0·34	0·37	0·41	0·44	0·48

$$\tau_{Rd} = (0\cdot25\, f_{ctk0\cdot05})/\gamma_c$$

where $f_{ctk0\cdot05}$ is the value in Table 4.1 and $\gamma_c = 1\cdot5$.

4.1.5. Bond strength

The basic bond strength of concrete for various values of f_{ck} may be obtained from Table 4.3. The values apply to 'good bond conditions' defined in Chapter 10.

The values in Table 4.3 have been derived from the following formulae using a value of $1\cdot5$ for γ_c

for plain bars $\qquad f_{bd} = (0\cdot36\,\sqrt{f_{ck}})/\gamma_c$

for high bend bars $\qquad f_{bd} = (2\cdot25\, f_{ctk0\cdot05})/\gamma_c$

The basic values should be modified when the reinforcement is in 'poor' bond conditions and when transverse pressures are present. For guidance on these, refer to Chapter 10.

MATERIALS AND DESIGN DATA

Table 4.3. Values of f_{bd} with $\gamma = 1.5$

f_{ck}	12	16	20	25	30	35	40	45	50
Plain bars	0.9	1.0	1.1	1.2	1.3	1.4	1.5	1.6	1.7
High bond bars where Dia. ≤ 32 mm or welded mesh fabrics made of ribbed wires	1.6	2.0	2.3	2.7	3.0	3.4	3.7	4.0	4.3

4.1.6. Deformation properties

The parameters required to determine the deformation of concrete structures include (a) strength of concrete, (b) properties of aggregates used in concrete, (c) details of concrete mix and (d) the environment surrounding the structure.

Clearly, when an accurate estimate of the deformation is sought, it is advisable to establish the values from appropriate data and conditions of use. EC2 gives values that may be used in the absence of more accurate data.

4.1.6.1. STRESS–STRAIN RELATIONSHIPS. Distinction should be made between the requirements of the analysis of the structure to obtain action effects (as in non-linear analysis) and the analysis of cross-sections to check the strength of chosen section sizes. For the latter EC2 provides three alternative stress strain diagrams: parabolic rectangular, bilinear and simplified rectangular (Fig. 4.2).

The σ_c–ϵ_c relationship shown in Fig. 4.2 for short-term loading can be expressed by the function

$$\frac{\sigma_c}{f_c} = \frac{kn - n^2}{1 + (k - 2)n}$$

where $n = \epsilon_c/\epsilon_{c1}$ (E_c and ϵ_{c1} values are both < 0), $\epsilon_{c1} = -0.0022$ (strain of the peak compressive stress f_c), $k = (1.1 E_{c,nom}) \epsilon_{c1}/f_c$ (f_c introduced as $-f_c$) and $E_{c,nom}$ denotes either the mean value E_{cm} of the longitudinal modulus of deformation (Table 4.4) or the corresponding design value E_{cd}.

4.1.6.2. MODULUS OF ELASTICITY E_c. The mean value of the secant modulus E_{cm} may be obtained from Table 4.4. The values are defined by $\sigma_c = 0$ and $\sigma_c = 0.4 f_{ck}$.

The values in the Table 4.4 have been derived from $E_{cm} = 9.5(f_{ck} + 8)^{0.33}$, where f_{ck} is in N/mm². Table 4.4 can also be used to determine the approximate value of E_{cm} at age t (other than 28 days), using the strength f_{ck} of concrete at any time t.

4.1.6.3. POISSON'S RATIO. When cracking is permitted for concrete in tension Poisson's ratio may be assumed to be zero; otherwise it may be taken as 0.2.

4.1.6.4. CREEP AND SHRINKAGE. Creep and shrinkage are mainly influenced by (a) ambient humidity, (b) the dimensions of the element and (c) the composition of the concrete. Creep is also influenced by the age at which load is first applied and the duration and magnitude of the loading. The final creep coefficients are given in Table 4.5; the final shrinkage strains are given in Table 4.6.

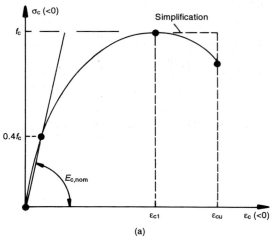

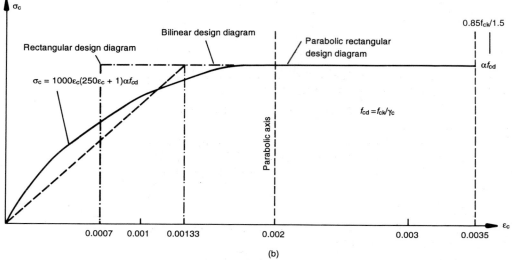

Fig. 4.2. (a) Stress—strain diagram for concrete for structural analysis; (b) design stress—strain diagram for concrete in compression

Table 4.4. Values of E_{cm}

Strength class C	12/15	16/20	20/25	25/30	30/37	35/45	40/50	45/55	50/60
E_{cm} kN/mm²	26·0	27·5	29·0	30·5	32·0	33·5	35·0	36·0	37·0

Table 4.5. Final creep coefficient of normal weight concrete

| Age at loading t_0: days | Notional size $2A_c/u$: mm | | | | | |
| | Dry conditions (inside, relative humidity = 50%) | | | Humid conditions (outside, relative humidity = 80%) | | |
	50	150	600	50	150	600
1	5·5	4·6	3·7	3·6	3·2	2·9
7	3·9	3·1	2·6	2·6	2·3	2·0
28	3·0	2·5	2·0	1·9	1·7	1·5
90	2·4	2·0	1·6	1·5	1·4	1·2
365	1·8	1·5	1·2	1·2	1·0	1·0

A_c = cross-sectional area of concrete, u = perimeter of that area. Table valid between −20°C and 40°C.

MATERIALS AND DESIGN DATA

Table 4.6. Final shrinkage strains of normal weight concrete: per mille

Location of the member	Typical relative humidity: %	Notional size $2A_c/u$: mm	
		≤ 150	600
Inside	50	−0·60	−0·50
Outside	80	−0·33	−0·28

A_c = cross-sectional area of concrete, u = perimeter of that area. Table valid between −20°C and 40°C.

Table 4.7. Slump classes

Class	Slump: mm
S1	10–40
S2	50–90
S3	100–150
S4	≥ 160

The values for shrinkage strain given in Table 4.6 relate to concrete having a plastic consistence of classes S2 and S3 as specified in ENV 206, clause 7.2.1. For concrete of classes S1 and S4 the values given in the Table 4.6 should be multiplied by 0·7 and 1·2 respectively.

Table 4.7 reproduces the details from ENV 206 with regard to slump classes. For classes S2 and S3 the slump may vary between 50 mm and 150 mm. It is not logical that mixes with this variation of slump, and hence w/c ratio, should have the same value of shrinkage strain.

It is more reasonable to assume that

(a) the values in Table 4.6 relate to an average value of slump for classes S2 and S3, i.e. an average slump of ~ 100 mm
(b) the multiplier 0·7 applies to a concrete with a slump of 40 mm
(c) the multiplier 1·2 applies to a concrete with a slump of 160 mm
(d) interpolation between these figures is permissible.

4.1.6.5. COEFFICIENT OF THERMAL EXPANSION. The coefficient of thermal expansion may be generally taken as $10 \times 10^{-6}/°C$.

4.2. Reinforcement

4.2.1. General
EC2 will rely on EN10 080 to prescribe the methods of production, details of classification for products, methods of testing, and methods of attestation of conformity. EN10 080 is at present available only as a pre-standard.

Products are classified according to their

- grade, denoting the specified characteristic yield stress f_{yk}
- ductility class
- size
- surface characteristics (ribs or indentations)
- weldability.

In this manual, only strength and ductility characteristics are discussed.

4.2.2. Strength (f_{yk})

As stated above, the grade of reinforcement steel denotes the specified characteristic yield stress f_{yk}. The latter is obtained by dividing the characteristic yield load by the nominal cross-sectional area of the bar. For products without a pronounced yield stress, the 0·2% proof stress $f_{0·2k}$ is substituted as the yield stress. See Figs 4.3 and 4.4 for typical and idealized stress–strain diagrams.

In general, ductility is inversely related to yield stress. Therefore in applications where ductility is critical (e.g. seismic design), it is important to ensure that the actual yield strength does not exceed the specified value by a large margin. Limits for the ratio of the actual yield stress to the specified strength are likely to be specified in EN10 080 in due course when seismic grade reinforcement is considered. At present, no UK standard gives any guidance.

4.2.3. Ductility

Ductility is defined using the strain at maximum load ϵ_{uk} and the ratio of the

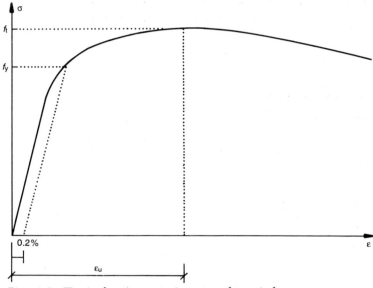

Fig. 4.3. Typical stress–strain curve for reinforcement

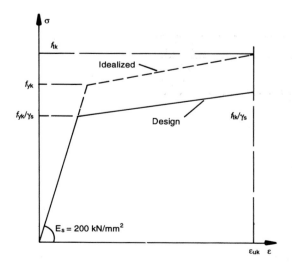

Fig. 4.4. Idealized stress–strain diagram for reinforcement

maximum and yield (strengths $f_t/f_y)_k$. Ductility is an essential property if advantage is to be taken of the plastic behaviour of structures. The greater the ductility, the greater will be the elongation in axially loaded members and the rotation capacity in members subjected to flexure. In members where the ultimate strength is governed by the yielding of reinforcement (shallow members with a low percentage of steel), clearly, the higher the value of ϵ_{uk}, the greater will be the ductility (i.e. the plateau of the stress–strain diagram is long).

When the ultimate strength is controlled by the strain in concrete reaching the limiting value, the length of the plastic zone influences the rotation. The greater this length, the greater will be the rotation. (f_t/f_y) indicates the length over which yield takes place, i.e. the length of the plastic zone (see Fig. 4.5).

Two classes of ductility are recognized in EC2. The requirements for these classes are given in Table 4.8.

All the values in Table 4.8 are boxed. Research is in progress across Europe to determine the ductility necessary in structures. As a result the values may be changed. Requirements for seismic design have not yet been considered.

In practice, in the UK, reinforcement size 16 and above will comply with the high ductility requirements. Smooth wire mesh and bar sizes 12 and less should be treated as possessing only normal ductility at present.

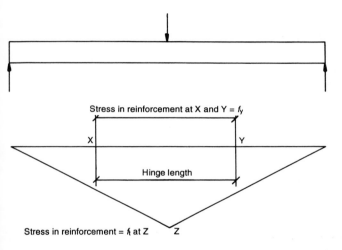

Fig. 4.5. Plastic zone

Table 4.8. Ductility requirement for reinforcement

	ϵ_{uk}	$(f_t/f_y)_k$
High ductility	>5%	>1·08
Normal ductility	>2·5%	>1·05

CHAPTER 5

Design of sections for bending and axial force

This chapter is concerned with the design for the ultimate limit state of sections subject to pure flexure, such as beams or slabs, sections subject to combined bending and axial load, such as columns, and prestressed sections.

The material covered in this chapter is covered in EC2 in the sections on

design stress—strain curves for reinforcement *(clause 4.2.2.3.2)*   *Clause 4.2.2.3.2*
design stress—strain curves for prestressing steel *(clause 4.2.3.3.3)*   *Clause 4.2.3.3.3*
design stress—strain curves for concrete *(clause 4.2.1.3.3)*   *Clause 4.2.1.3.3*
basic assumptions for section design *(clause 4.3.1.2)*   *Clause 4.3.1.2*
brittleness and hyperstrength *(clause 4.3.1.3)*   *Clause 4.3.1.3*
limitations imposed by redistribution *(clause 2.5.3.4.2)*.   *Clause 2.5.3.4.2*

5.1. Basic assumptions

The basic assumptions about section behaviour are very similar to those adopted by many, if not most, modern codes of practice. The formulation used in EC2 is taken from the CEB Model Code.[8] They differ in detail from those in the UK Code, BS8110, in ways that, as will be seen, make calculations more complex in some cases, but the practical outcome is not significantly different. The assumptions define the stress—strain responses to be assumed for steel and concrete and the assumptions to be made about the strains at the ultimate limit state. It is these assumptions about the strain that define failure.

5.1.1. Stress—strain curves

The information required to obtain the design stress—strain curves for concrete, ordinary reinforcement and prestressing steel is given in *clauses 4.2.1, 4.2.2 and 4.2.3* respectively.

For reinforcement and prestressing steel, EC2 specifies the use of bilinear stress—strain curves. These are shown in *clauses 4.2.2.3.2 and 4.2.3.3.3* for reinforcing and prestressing steels respectively. In each case, one can choose between two possible bilinear diagrams for the design of sections: one with a horizontal top branch and one with an inclined top branch. These curves are shown in Fig. 5.1. The inclined upper branches terminate at a strain of 0·01 and a stress of f_{tk}, the tensile strength of the steel. *Clause 3.4.2* gives f_{tk}/f_{yk} values of 1·08 for high-ductility steel and 1·05 for normal-ductility steel. Clearly, use of the curves with

Clause 4.2.1
Clause 4.2.2
Clause 4.2.3
Clause 4.2.2.3.2
Clause 4.2.3.3.3

Clause 3.4.2

DESIGN OF SECTIONS FOR BENDING AND AXIAL FORCE

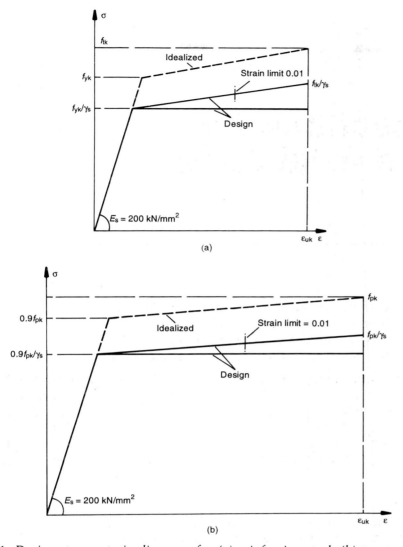

Fig. 5.1. Design stress–strain diagrams for (a) reinforcing steel, (b) prestressing steel

inclined top branches will give some economic advantage over use of the horizontal top branch. Potentially, this advantage could give an 8% saving in reinforcement, but a saving approaching this value will rarely be achievable and the calculations involved will be considerably more complex. The full saving will be available only where it is certain at the design stage that a high-ductility steel will be used, and where the neutral axis depth at the ultimate limit state will be around $0 \cdot 25$. Above this the steel stress will be below f_{tk}/α_m, while for lower neutral axis depths the effect of limiting the tension strain to $0 \cdot 01$ will be a lower average stress in the compression zone and hence a deeper neutral axis than if the strain were unlimited. Furthermore, the requirement to limit the strain to $0 \cdot 1$ leads to restrictions on the amount of redistribution that can be carried out. For example, the value of $M/bd^2 f_{ck}$ corresponding to a neutral axis depth of $0 \cdot 11d$, the limiting depth for 30% redistribution for concrete grades above C35/45, is $0 \cdot 05$ where the horizontal upper branch is used and $0 \cdot 031$ where the inclined upper branch is used. Compression steel would therefore be required significantly earlier. Thus, the added complexity involved in using the inclined upper branch does not generally seem worthwhile, and the horizontal upper branch will be used in the derivation of all equations, charts and tables in this chapter.

For concrete, three possibilities are described in *clause 4.2.1.3.3(b)*. The preferred idealization is the parabolic–rectangular diagram, but a bilinear diagram and a rectangular diagram are also permitted. These three diagrams are compared in Fig. 5.2.

Clause 4.2.1.3.3(b)

The maximum stress in the characteristic diagram is given by the characteristic concrete strength multiplied by a coefficient α. This coefficient is described in *clause 4.2.1.3.3* as '... taking account of long term effects on the compressive strength and of unfavourable effects resulting from the way the load is applied'. A value of 0·85 is suggested generally, reducing to 0·8 if the compression zone is decreasing in width towards the extreme compression fibre. It is arguable, however, that the reason for the introduction of this factor is more related to the idealization of the shape of the stress–strain diagram than to the nature of the loading. Fig. 5.3 is an attempt to illustrate this. For the areas under the actual curve and the idealized curve to be the same, the maximum stress level for the idealized curve must be below the maximum stress of the 'true' diagram.

Clause 4.2.1.3.3

Table 5.1 compares the three permitted idealizations in terms of the average stress over a rectangular compression zone and the distance from the compression

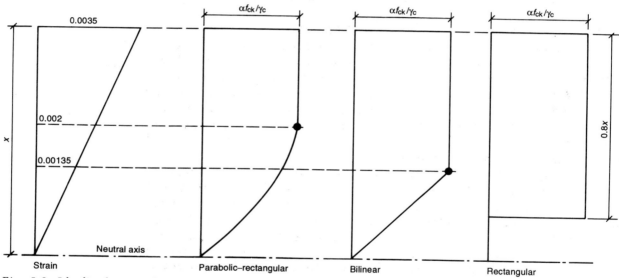

Fig. 5.2. Idealized stress distributions

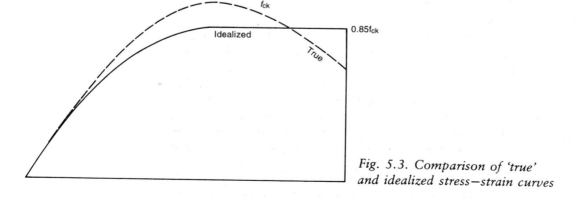

Fig. 5.3. Comparison of 'true' and idealized stress–strain curves

Table 5.1. Comparison of stress block parameters

Grade	Parabolic–rectangular		Bilinear		Rectangular	
	Average stress	Centroid	Average stress	Centroid	Average stress	Centroid
12	5·51	0·416	5·49	0·411	5·44	0·40
16	7·34	0·416	7·32	0·411	7·25	0·40
20	9·18	0·416	9·15	0·411	9·07	0·40
25	11·47	0·416	11·44	0·411	11·33	0·40
30	13·76	0·416	13·72	0·411	13·60	0·40
35	16·06	0·416	16·01	0·411	15·87	0·40
40	18·35	0·416	18·30	0·411	18·13	0·40
45	20·64	0·416	20·58	0·411	20·40	0·40
50	22·94	0·416	22·87	0·411	22·67	0·40

face of the section to the centre of compression. As well as providing a convenient comparison of the idealizations, Table 5.1 provides information in a useful form for design calculations.

It will be seen from Table 5.1 that the results obtained from the three idealizations will be so similar as to be indistinguishable for all normal purposes.

5.1.2. Assumptions relating to the strains at ultimate

The basic assumptions here are that plane sections remain plane and the strain in reinforcement is the same as the strain in the concrete at the same level. These assumptions are universally accepted for the design of members containing bonded ordinary reinforcement. For prestressed sections, allowance has to be made for the strain in the steel prior to its being bonded to the concrete. After bonding, the change in strain in the steel is assumed to be the same as the change in strain in the concrete. This assumption is of course invalid for beams with unbonded tendons. These lie outside the scope of EC2: Part 1.

The assumptions are not strictly true. The deformations within a section are very complex and, locally, plane sections do not remain plane. Nor, due to local bond slip, are the strains in the concrete exactly the same as those in the steel. Nevertheless, on average, the assumptions are correct and are certainly sufficiently true for practical purposes for the design of normal members. One area where they are not adequate is in short members. For this reason, deep beams are not designed using the provisions of the chapter, and nor are members such as corbels.

No absolute limit is given in EC2 for the maximum tensile strain in the reinforcement unless a steel stress–strain curve with an inclined upper branch is being used. However, there is clearly a limiting strain that defines the failure of any particular type of reinforcement. This is the strain corresponding to the maximum stress that the steel will withstand. It therefore seems logical to limit the tensile strain in reinforcement to the characteristic maximum strain value given for the particular class of reinforcement being used. This is 2·5% for normal-ductility steel and 5% for high-ductility steel. In practice, some codes (e.g. BS8110) give no limit, while others (e.g. the CEB Model Code and the DIN Code) give a limit of 1%. The question of whether a limit should be specified in EC2 was the subject of extensive debate within the Editing Group. The final conclusion reached was that where a stress–strain response was chosen for a steel that had a rising stress beyond yield, a strain limit of 1% should be applied. If a perfectly plastic response was assumed after yield, it was not necessary to impose a limit but a limit could be applied if this proved convenient. Possible situations where this may help are considered below. The lack of a limit may seem illogical, as there is undoubtedly a strain beyond which rupture will occur. However, in practice, the imposition of a limit has minimal effect on the design of a section beyond making the calculation more involved.

It is universal to define failure of concrete in compression by means of a limiting compressive strain. The formulation of the limit varies from code to code, for example the American Concrete Institute Code, ACI 318, uses a limit of 0·003 while BS8110 uses 0·0035. EC2 adopts values taken from the CEB Model Code. These comprise a limit of 0·0035 for flexure and for combined bending and axial load where the neutral axis remains within the section, and a limit of between 0·0035 and 0·002 for sections loaded so that the whole section is in compression. This is illustrated in Fig. 5.4.

The logic behind the reduction in the strain limit is that in axial compression, failure will occur at the strain corresponding to the attainment of the maximum compressive stress. This is 0·002. In flexure, considerably higher strains can be reached before the maximum capacity of the section is reached and the value of 0·0035 has been obtained empirically. A means is needed to interpolate between the values of 0·0035 for flexure and 0·002 for axial load; Fig. 5.4 provides this.

It is of interest to investigate the sensitivity of the calculated section capacity to the assumptions made about the ultimate strains. In Table 5.2 the effect of limiting the steel strain to 0·01 on the calculated steel percentage and neutral axis depth at the ultimate limit state can be seen. The calculations have been performed for grade C25/30 concrete and 460 N/mm² reinforcement.

Table 5.3 compares the values of M/bd^2 calculated applying the EC2 compressive strain limits for situations where the whole section is in compression with those calculated assuming a maximum strain of 0·0035 regardless of the neutral axis position. The calculations have been carried out for a symmetrically reinforced rectangular column reinforced with 2% of 460 grade steel. The concrete grade has been taken as C25/30.

It will be seen that the calculated strengths are remarkably insensitive to variations in the assumptions. This supports the case for adopting the simplest possible formulation, even though a more complex formulation may be logically

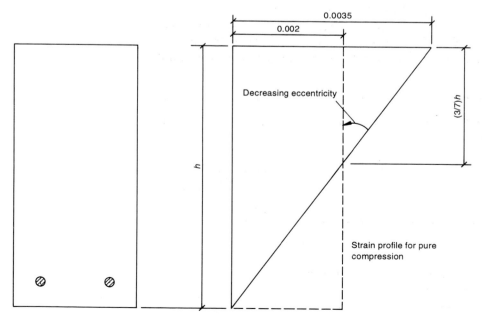

Fig. 5.4. *Ultimate strains in member wholly in compression*

DESIGN OF SECTIONS FOR BENDING AND AXIAL FORCE

Table 5.2. Comparison of calculated steel percentages and neutral axis depths where: (a) the strain is limited to 0·01 in tension; (b) there is no limit on the tension strain

M/bd^2	Reinforcement ratio		n-axis depth	
	(a)	(b)	(a)	(b)
2·649	0·742	0·742	0·2	0·259
2·5	0·695	0·695	0·2	0·243
2	0·544	0·543	0·2	0·19
1·5	0·40	0·398	0·1	0·139
1	0·264	0·26	0·1	0·091
0·5	0·13	0·128	0·1	0·045
0	0	0	0	0

Table 5.3. Comparison of moment capacities of columns where: (a) the compressive strain obeys the EC2 limits; (b) the limiting compressive strain is 0·0035

N/bh	M/bh^2		x/h	
	(a)	(b)	(a)	(b)
16	2·34	2·34	0·99	0·99
17	1·97	1·98	1·06	1·05
18	1·60	1·62	1·17	1·13
19	1·22	1·25	1·33	1·23
20	0·84	0·86	1·64	1·37
21	0·46	0·47	2·42	1·59

more defensible. In this respect, EC2 appears to have adopted an unnecessarily complex approach.

5.2. Design equations

5.2.1. Singly reinforced beams and slabs

The conditions in a singly reinforced rectangular section at the ultimate limit state are assumed to be as shown in Fig. 5.5. From Fig. 5.5, it will be seen that the following equations can be derived from consideration of equilibrium of axial forces and moments

$$f_{av}bx = f_{yd}A_s \text{ or } x/d = f_{yd}\rho/f_{av} \quad (5.1)$$

$$M = f_{av}bx(d-\beta x) \text{ or } M/bd^2 = f_{av}(1-\beta x/d)x/d \quad (5.2)$$

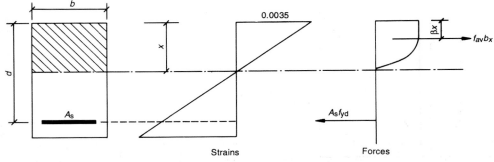

Fig. 5.5. *Conditions in a singly reinforced section at ultimate*

where A_s is the area of tension steel, f_{av} is the average stress over the compression zone, f_{yd} is the design yield strength of the steel, β is the ratio of the distance of the centre of compression from the compression face to the neutral axis depth, and ρ is the reinforcement ratio (A_s/bd).

For the parabolic–rectangular stress–strain curve, it can be derived by use of the partial safety factors given in EC2 that

$$f_{av} = 0\cdot 459 f_{ck}$$
$$f_{yd} = f_{yk}/1\cdot 15$$
$$\beta = 0\cdot 416$$

Substitution for x/d in equation (5.2) from equation (5.1), using the values given above and rearranging, gives

$$A_s = f_{ck}bd(0\cdot 633 - \sqrt{0\cdot 4 - 1\cdot 46K})/f_{yk} \qquad (5.3)$$

where $K = M/bd^2 f_{ck}$.

It may sometimes be useful to be able to obtain the neutral axis depth and the lever arm. These are given by

$$x/d = 1\cdot 895 f_{yk} A_s/(f_{ck}bd) \qquad (5.4)$$

$$z/d = 1 - 0\cdot 416 x/d \qquad (5.5)$$

It is now necessary to consider the limits to the use of singly reinforced sections. Clause 2.5.3.4.2 gives limits to the neutral axis depth at the ultimate limit state as a function of the amount of redistribution carried out in the analysis. These are also given here, for convenience.

Clause 2.5.3.4.2

For concrete of grade C35/45 or less

$$x/d \leq 0\cdot 8\delta - 0\cdot 35$$

and in no case is x/d greater than $0\cdot 45$. For concrete of grade greater than C35/45

$$x/d \leq 0\cdot 8\delta - 0\cdot 45$$

and in no case is x/d greater than $0\cdot 35$.

In the above relationships, δ is the ratio of the redistributed moment to the moment before redistribution. δ is limited as a function of the type of reinforcement used, as follows. For high-ductility steel, $\delta \geq 0\cdot 70$; for normal-ductility steel $\delta \geq 0\cdot 85$.

The neutral axis limits set out above can now be substituted into equation (5.2) to give limiting values of the parameterized moment K. This gives the following limits.

For grade C35/45 concrete or below

$$K_{lim} = (0\cdot 8\delta - 0\cdot 35)(0\cdot 525 - 0\cdot 153\delta) \leq 0\cdot 166 \qquad (5.6)$$

For concrete of grade greater than C35/45

$$K_{lim} = (0\cdot 8\delta - 0\cdot 45)(0\cdot 544 - 0\cdot 153\delta) \leq 0\cdot 1376 \qquad (5.7)$$

The expressions for the calculation of the areas of reinforcement and the limiting values of K can be presented as design charts. This is done in Fig. 5.6. The lower part of Fig. 5.6 gives the mechanical reinforcement ratio $A_s f_{yk}/bd f_{ck}$ in terms of the non-dimensional moment parameter $K = M/bd^2 f_{ck}$. The upper part of Fig. 5.6 gives values of the limiting neutral axis depth corresponding to values of the redistribution parameter δ. If one reads across to the appropriate line from the appropriate value of δ and then drops down to the curve, the corresponding value of K_{lim} can be read off the left-hand axis. This chart can now be used for the design of singly reinforced rectangular sections or flanged sections where the neutral

DESIGN OF SECTIONS FOR BENDING AND AXIAL FORCE

axis remains within the flange. The use of the chart is illustrated below by a simple example.

Example 5.1
A rectangular section 500 mm deep by 300 mm wide is to be designed to carry an ultimate moment of 120 kN m. The characteristic strength of the

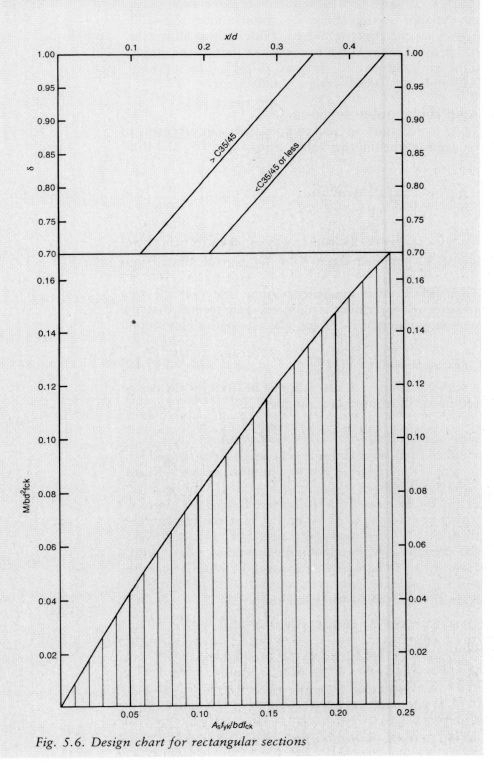

Fig. 5.6. *Design chart for rectangular sections*

reinforcement is 500 N/mm² and that of the concrete is 25 N/mm². 15% of redistribution was carried out in the analysis for the ultimate limit state.

Assume that the effective depth is 450 mm. This enables K to be calculated as $120 \times 10^6/(300 \times 450^2 \times 25) = 0.079$. From Fig. 5.6, the mechanical reinforcement ratio can be read off as 0.1, giving an area of reinforcement of $0.1 \times 300 \times 450 \times 25/500 = 675$ mm². A check should also be carried out to ensure that K is less than K_{lim}. Since 15% of redistribution has been done, $\delta = 0.85$ and hence, from the upper section of Fig. 5.6, the maximum allowable neutral axis depth is 0.33 times the effective depth. If one follows down the line for $x/d = 0.33$ until it touches the curve in the lower part of the graph, and then reads across to the left-hand axis, K_{lim} is given as 0.13. This is greater than 0.079, and hence a singly reinforced section is satisfactory.

5.2.2. Doubly reinforced rectangular sections

If K is greater than K_{lim}, it is necessary to add compression reinforcement to maintain the neutral axis depth at the limiting value. Equations (5.8) and (5.9) will achieve this

$$A_{sc} = \frac{(K - K_{lim})f_{ck}bd^2}{0.87f_{yk}(d - d')} \qquad (5.8)$$

$$A_s = A_{sc} + A_{s,lim} \qquad (5.9)$$

where $A_{s,lim}$ is the value of A_s obtained when $K = K_{lim}$ is substituted into equation (5.3).

The equations derived in this section implicitly assume first that the area displaced by the compression reinforcement is ignored, and second that the compression reinforcement is working at its design yield strength. This second assumption is true only if

$$f_{sd}/E_s < 0.0035(x - d')/x \qquad (5.10)$$

Fig. 5.6, in conjunction with equations 5.8 and 5.9, can be used for the design of doubly reinforced sections as illustrated in Example 5.2.

Example 5.2: Rectangular beam with compression reinforcement
The amounts of tension and compression reinforcement are to be calculated for a section 300 mm wide and of 450 mm overall depth to sustain a design ultimate moment of 255 kN m. The distance from the compression face to the centroid of any compression steel and the distance from the tension face to the centroid of the tension steel may both be assumed to be 50 mm. The reinforcement has a characteristic strength of 500 N/mm² and the concrete is grade C30/37. 15% of redistribution was carried out in the analysis for the ultimate limit state.

The value of δ corresponding to 15% redistribution is 0.85. Fig. 5.6 can be used to obtain the value of K_{lim} corresponding to this: it is 0.13. Equation (5.8) can now be used to obtain the area of compression steel

$$K = 255 \times 10^6/(400 \times 300 \times 30) = 0.177$$

Substitution into equation (5.8) gives

$$A_{sc} = \frac{(0.177 - 0.13) \times 30 \times 300 \times 400^2}{0.87 \times 500 \times (400 - 50)}$$

$$= 444 \text{ mm}^2$$

Fig. 5.6 gives a value of $A_s f_{yk}/(bdf_{ck})$ corresponding to $K_{lim} = 0.172$. This corresponds to a reinforcement area of 1238 mm². Substitution into equation (5.9) now gives the area of tension reinforcement as

$$A_s = 444 + 1238 = 1682 \text{ mm}^2$$

There are inconsistencies in the treatment of redistribution and the limits to neutral axis depth that may be of interest.

The first of these is related to plastic design. The maximum redistribution allowed in beams is 30% (i.e. $\delta = 0.7$). For C40/50 concrete, this corresponds to a neutral axis depth of $0.112d$. However, for slabs, *clause 2.5.3.5* states that plastic design may be used for any section where the neutral axis depth is less than $0.25d$. The limiting moment for the use of plastic design for slabs is thus $K_{\text{lim}} = 0.103$ for all conditions. Below this value, unlimited redistribution is permitted. For the plastic analysis of beams, *clause 2.5.3.4.4* refers the reader to *Appendix 2*. This in turn states that the rules in *clause 2.5.3.5* may be applied. Thus the same rules on limiting neutral axis depth for the application of plastic design apply to beams as apply to slabs, and unlimited redistribution can apparently be carried out in beams if K is less than 0.103.

Clause 2.5.3.5

Clause 2.5.3.4.4
Appendix 2
Clause 2.5.3.5

The second point to note is that less redistribution is permitted with higher strength concrete. Thus, given that a design using C35/45 concrete just meets the limiting neutral axis depth for the amount of redistribution carried out, if the member were constructed using concrete of a higher characteristic strength than that specified it would be unsafe. Clearly, this is a possibility that the designer must simply ignore.

5.2.3. Design of flanged sections (T- or L-beams)

Provided that the neutral axis at the ultimate limit state remains within the flange of a T- or L-beam, the section may be treated as a rectangular section and the equations derived above may be used. In fact, it will be satisfactory to consider the section as rectangular for neutral axis depths up to 1.25 times the flange depth. This can be seen from consideration of the simplied rectangular stress block shown in Fig. 5.2. From experience, it will only very infrequently be necessary to do other than consider a flanged beam as rectangular. If the neutral axis does need to be deeper than 1.25 times the flange depth, then it will be sufficiently accurate to assume that the whole of the outstanding parts of the flanges is carrying a concrete stress of $0.85f_{cd}$. This leads to the equilibrium equations

$$A_s f_{yd} = 0.85 f_{cd} h_f (b - b_r) + 0.459 f_{ck} b_r x \qquad (5.11)$$
$$M = 0.459 f_{av} b_r x (d - 0.416x) + 0.85 f_{cd} h_f (b - b_r)(d - h_f/2) \qquad (5.12)$$

where b is the overall breadth of the section, b_r is the breadth of the rib and h_f is the depth of the flange.

These equations are too complicated to be worth refining further. They can be used either as a basis for deriving design charts or iteratively for direct design. A suitable iterative approach would be as follows

(a) guess a value of x
(b) substitute for x in equation (5.12) and calculate M
(c) if M is not close to the required value, return to step (a)
(d) substitute x into equation (5.11) to obtain a steel area.

An alternative simplified approach for flanged beams where the neutral axis exceeds 1.25 times the flange depth is to assume that the centre of compression is at mid-depth of the flange. This enables a reinforcement area to be obtained from

$$A_s = M/[f_{yd}(d - h_f/2)] \tag{5.13}$$

This simplified formula will be sufficiently accurate for all normal purposes, but does not provide any means of checking the neutral axis depth. This could be checked by use of equation (5.11) if it were felt that the limit was likely to be approached.

5.2.4. Design of rectangular column sections

The basic equations for equilibrium of a rectangular section subjected to combined bending and axial load for situations where the neutral axis remains within the section are

$$N_{Rd} = f_{av}bx + \Sigma f_s A_s \tag{5.14}$$

$$M_{Rd} = f_{av}bx(h/2 - \beta x) + \Sigma f_s A_s(h/2 - d_i) \tag{5.15}$$

In equation (5.15), moments have been taken about the centroid of the concrete section. The summation signs indicate a summation over all the levels of reinforcement within the section. In carrying out the summation, tensile stresses must be taken as negative. d_i is the distance from the compressive face of the section to the ith layer of reinforcement.

Assuming that the values for the partial safety factors on the steel and concrete are taken as the indicative values given in EC2, $0 \cdot 459 f_{ck}$ can be substituted for f_{av} and $0 \cdot 416$ for β. The resulting equations are rigorous for situations where the neutral axis remains within the section. Where the whole section is in compression, more complex expressions are required to deal with (a) the portion of the parabolic curve cut off by the bottom of the section, and (b) the reduction in the ultimate strain at the compressive face according to Fig. 5.4. One approach is to calculate the concrete force on the basis of the full neutral axis depth and then deduct the force and moment corresponding to the area below the bottom of the section. The average stress over the whole depth x is given by

$$f_{av} = 0 \cdot 5667 f_{ck}(1 - \beta/3) \tag{5.16}$$

$$c = h/2 - x(\beta^2 - 4\beta + 6)/(12 - 4\beta) \tag{5.17}$$

where $\beta = 0 \cdot 002/\epsilon_u$.

If the whole section is in compression

$$\epsilon_u = 0 \cdot 002 x/(x - 3h/7) \tag{5.18}$$

The average stress on the area beyond the section is given by

$$f'_{av} = 0 \cdot 5667 f_{ck} \alpha(1 - \alpha/3) \tag{5.19}$$

The distance from the centre of the section to the centroid of the compression force outside the section is given by

$$c' = x - (h/2) - (x - h)(8 - 3\alpha)/(12 - 4\alpha) \tag{5.20}$$

In the equations (5.19) and (5.20), α is $\epsilon_b/0 \cdot 002$, where ϵ_b is the strain at the bottom of the section.

The compressive force provided by the concrete is now given by

$$C_c = f_{av}bx - f'_{av}(x - h) \tag{5.21}$$

and the concrete contribution to the moment is given by

$$M_c = f_{av}bxc + f'_{av}b(x - h)c' \tag{5.22}$$

DESIGN OF SECTIONS FOR BENDING AND AXIAL FORCE

None of these equations are convenient for use for the design of column sections by hand, but they can easily be used to construct design charts. Figs 5.7–5.18 are a series of parameterized charts for the design of symmetrically reinforced rectangular columns derived using these equations. Three sets of charts are included

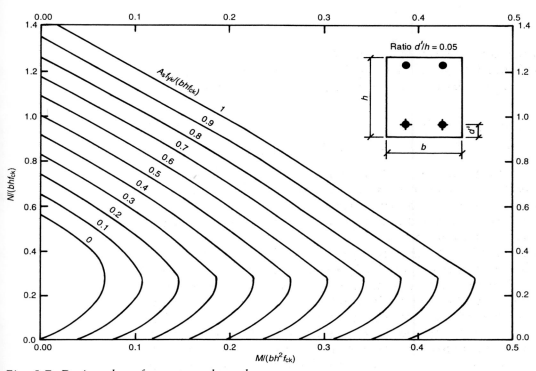

Fig. 5.7. Design chart for rectangular columns

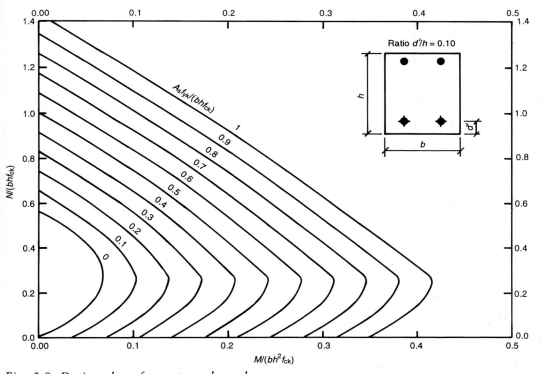

Fig. 5.8. Design chart for rectangular columns

here. The first set (Figs 5.7–5.10) can be used for sections where either the reinforcement is concentrated at the corners, as shown in the insets, or the steel is distributed along the sides parallel to the axis of bending. The second set of charts (Figs 5.11–5.14) is for sections where the reinforcement is distributed in

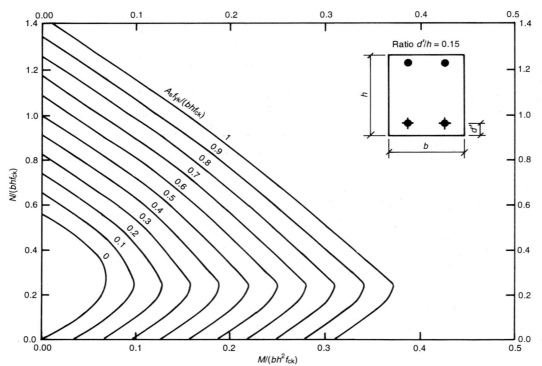

Fig. 5.9. Design chart for rectangular columns

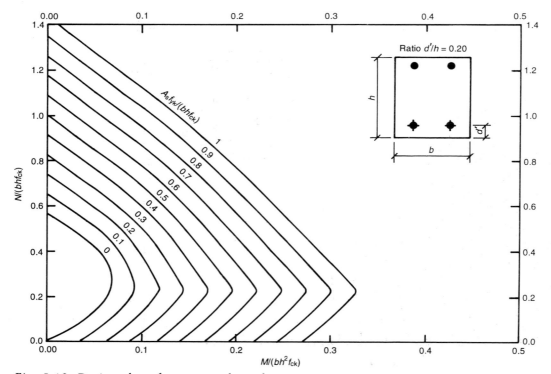

Fig. 5.10. Design chart for rectangular columns

the most disadvantageous way along the sides of the section that are perpendicular to the axis of bending. The final set of charts (Figs 5.15–5.18) is for sections where reinforcement is distributed evenly around the section perimeter. The use of these charts is illustrated by Example 5.3 below.

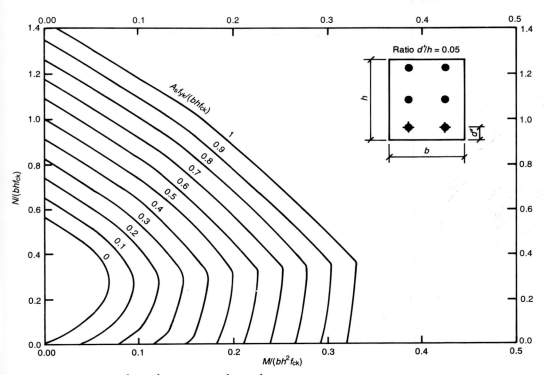

Fig. 5.11. Design chart for rectangular columns

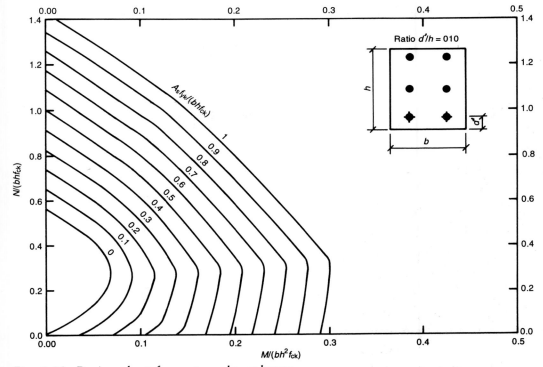

Fig. 5.12. Design chart for rectangular columns

Example 5.3
A rectangular column section 500 mm deep and 300 mm wide is to be designed to carry a design ultimate axial load of 1875 kN and a design ultimate moment of 280 kN m. The characteristic strength of the reinforcement is 500 N/mm^2 and that of the concrete is 30 N/mm^2.

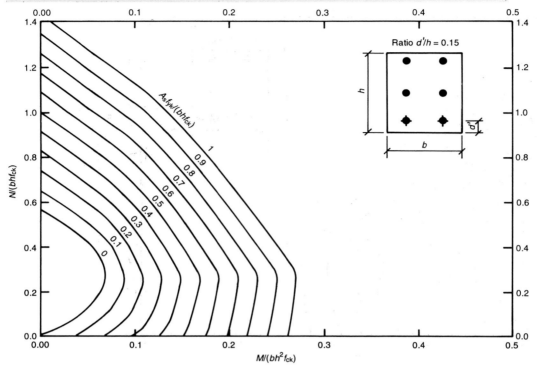

Fig. 5.13. *Design chart for rectangular columns*

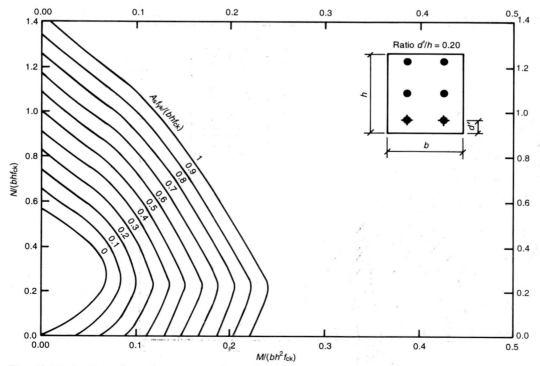

Fig. 5.14. *Design chart for rectangular columns*

The non-dimensional moment and axial load parameters are calculated as follows

$$N/bhf_{ck} = 1875 \times 1000/(300 \times 500 \times 30) = 0.417$$
$$M/bh^2 f_{ck} = 280 \times 10^6/(300 \times 500^2 \times 30) = 0.124$$

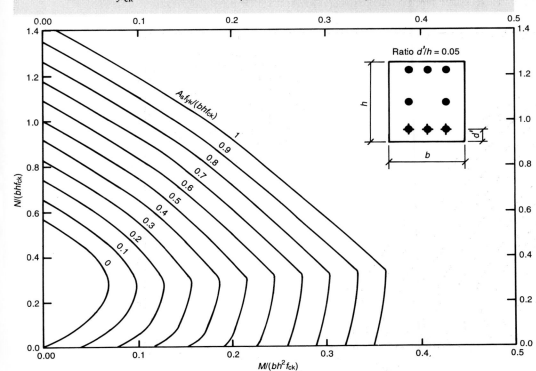

Fig. 5.15. Design chart for rectangular columns

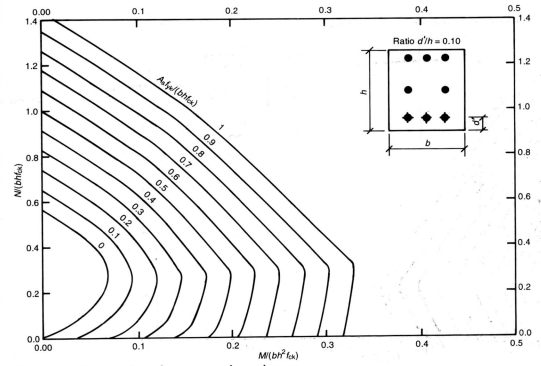

Fig. 5.16. Design chart for rectangular columns

Assuming that the centroid of the steel in each face is 50 mm from the face and that the reinforcement is concentrated at the corners of the section, Fig. 5.7 can be used to establish that the mechanical reinforcement ratio is 0·27. This gives a total steel area of 0·27 × (30/500) × 500 × 300 = 2430 mm².

Charts can also be produced for circular column sections: a set of parameterized

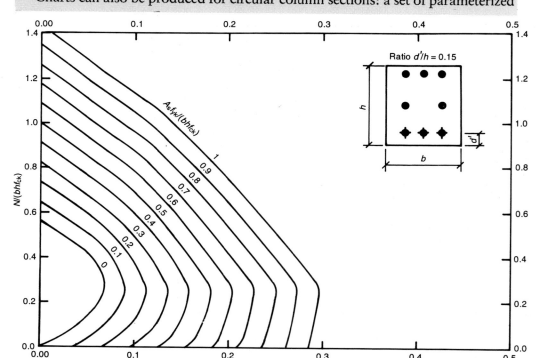

Fig. 5.17. *Design chart for rectangular columns*

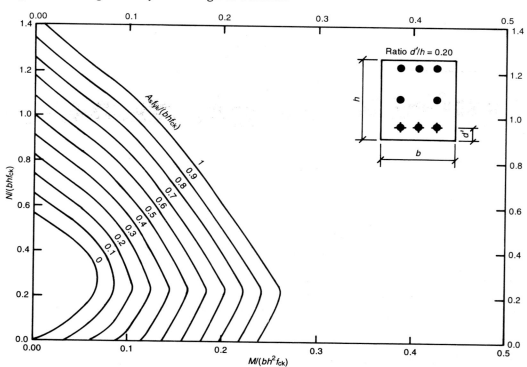

Fig. 5.18. *Design chart for rectangular columns*

charts is shown in Figs 5.19–5.22. The charts are drawn assuming that the section contains six reinforcing bars, which is the minimum that can reasonably be used in a circular section. It is found that there is no single arrangement of the reinforcement relative to the axis of bending that will always give the minimum

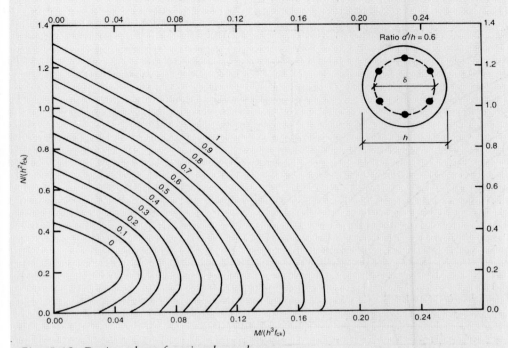

Fig. 5.19. *Design chart for circular columns*

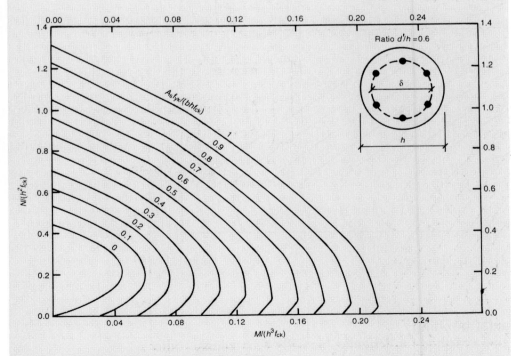

Fig. 5.20. *Design chart for circular columns*

strength. The charts are therefore drawn to give a lower-bound envelope to the interaction diagrams for various arrangements of the bars.

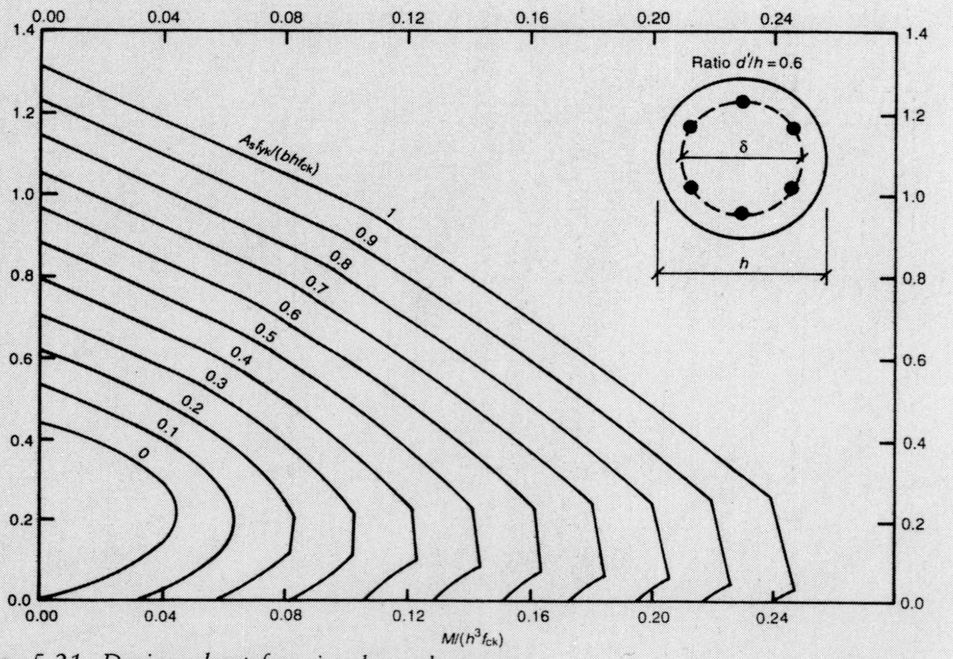

Fig. 5.21. Design chart for circular columns

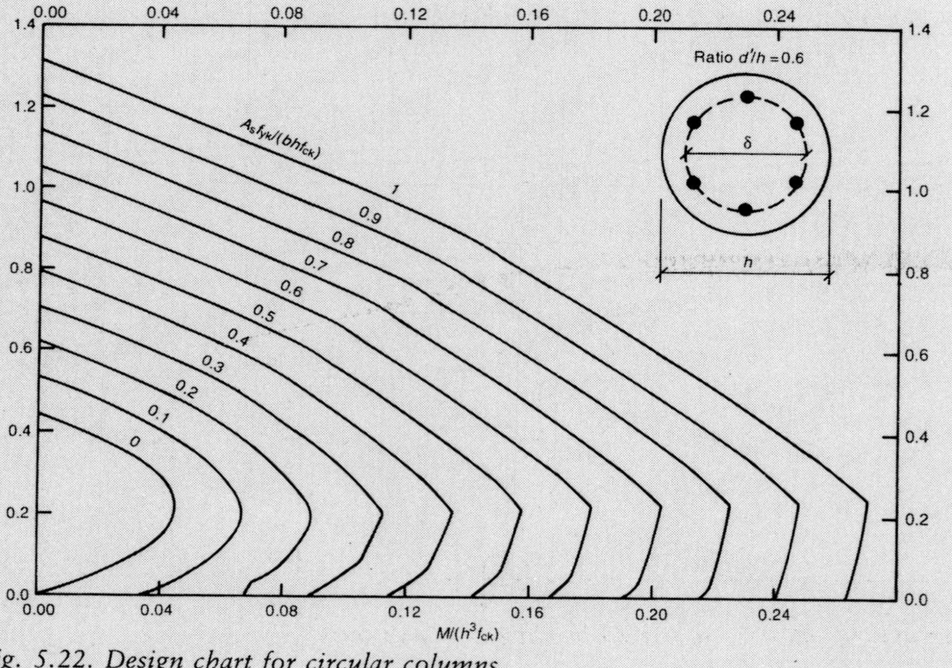

Fig. 5.22. Design chart for circular columns

5.2.5. Design for biaxial bending
EC2 does not directly give a method for designing biaxially bent columns other than working from first principles. This is not easy to do without design charts or a suitable computer program. In this section, a number of simplified methods

DESIGN OF SECTIONS FOR BENDING AND AXIAL FORCE

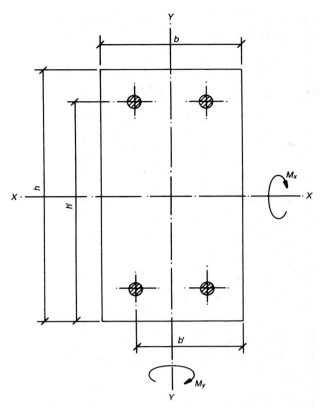

Fig. 5.23. Notation used in BS8110 method for biaxially bent columns

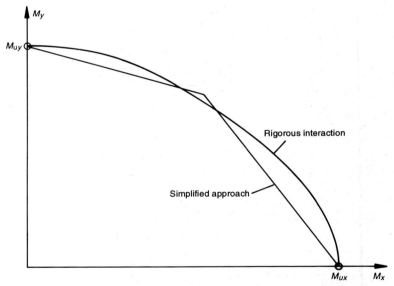

Fig. 5.24. Comparison of simplified approach to biaxial bending with rigorous analysis

are considered that will allow the design of rectangular sections to be carried out relatively simply. In addition, a set of design interaction charts is produced for the design of symmetrically reinforced biaxially bent rectangular sections. These are shown in Figs 5.26–5.35.

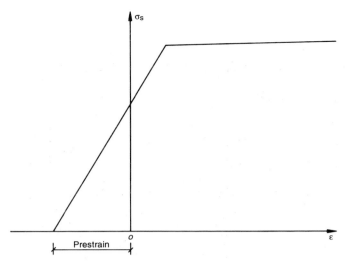

Fig. 5.25. *Effective steel stress–strain curve for design of prestressed sections at ultimate*

Three methods of dealing with biaxially bent rectangular sections are presented here. The methods are given in order of increasing simplicity and increasing approximation.

The first is an approach used in the old UK code for structural concrete, CP110. This is based on the observation that the form of the $M_x - M_y$ interaction diagram can conveniently be represented by a super-ellipse, which has an equation of the form

$$x^a + y^a = k$$

If $a = 2$, this equation describes a circle; if $a = 1$ it describes a straight line. At loads approaching the squash load, the $M_x - M_y$ interaction diagram approaches a circle, while in the region of the balance point it is close to a straight line. CP110 adopted equation (5.23) as a means of describing the complete interaction surface

$$(M_x/M_{ux})^a + (M_y/M_{uy})^a = 1 \qquad (5.23)$$

A convenient parameter for defining the proximity to the squash load is the ratio N/N_{uz}, and CP110 assumed a linear relation between this parameter and the exponent a. The CP110 values are given in Table 5.4.

A study[9] has shown that this approach is good, and generally slightly on the conservative side for sections where the steel is concentrated in the corners. It is more conservative where the reinforcement is distributed fairly uniformly around the perimeter of the section.

The difficulty with the approach from the practical point of view is that it cannot be used as a direct design method, since N_{uz} can be established only once the reinforcement area has been found. It therefore has to be used iteratively. An initial

Table 5.4. CP110 parameters

N/N_{uz}	a
0·2	1·0
0·4	1·33
0·6	1·66
0·8	2·0

DESIGN OF SECTIONS FOR BENDING AND AXIAL FORCE

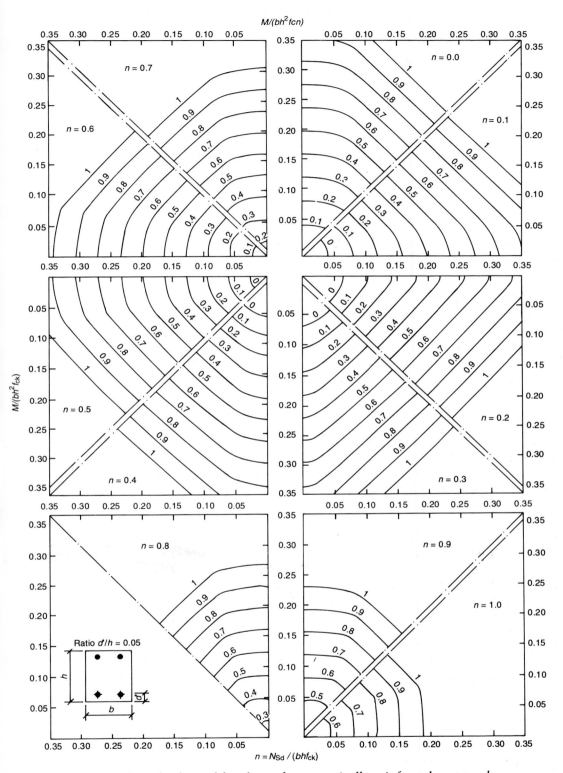

Fig. 5.26. Design chart for biaxial bending of symmetrically reinforced rectangular sections

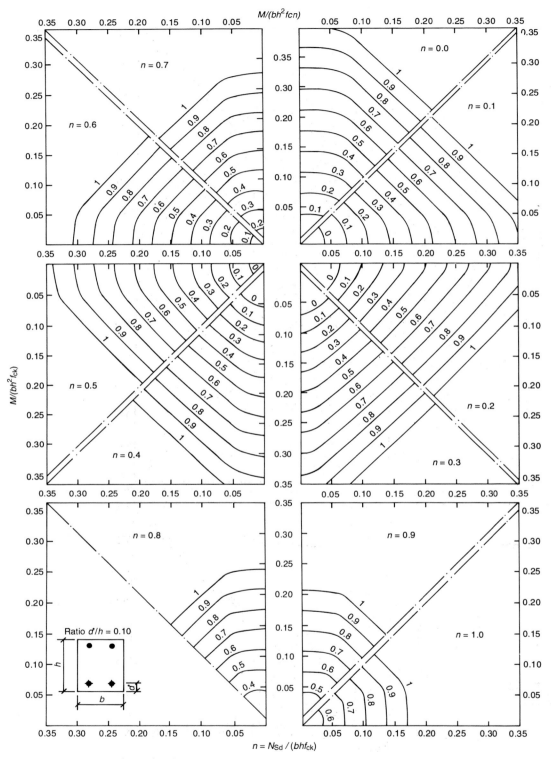

Fig. 5.27. Design chart for biaxial bending of symmetrically reinforced rectangular sections

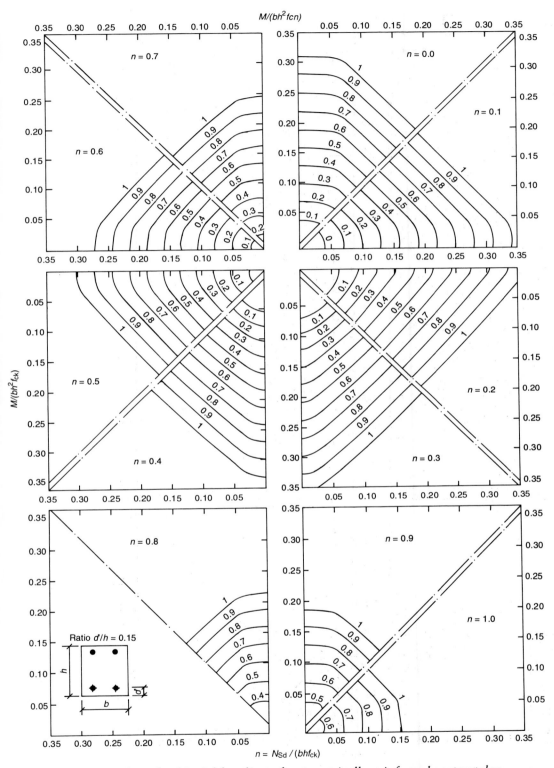

Fig. 5.28. Design chart for biaxial bending of symmetrically reinforced rectangular sections

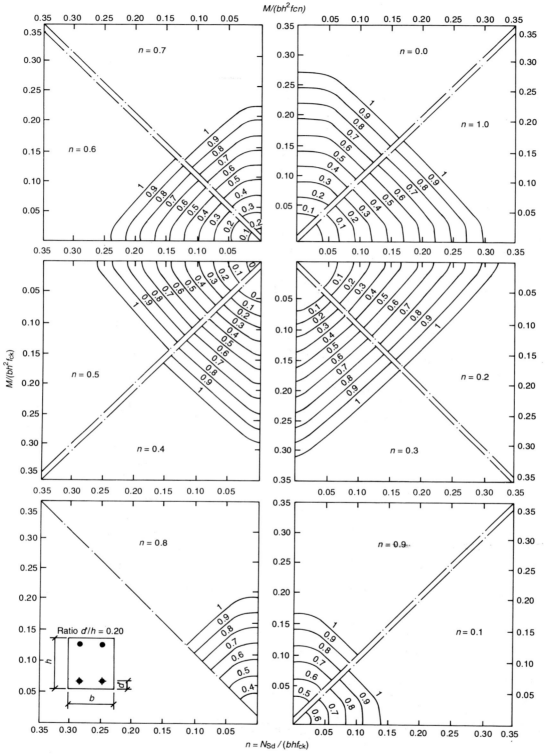

Fig. 5.29. *Design chart for biaxial bending of symmetrically reinforced rectangular sections*

DESIGN OF SECTIONS FOR BENDING AND AXIAL FORCE

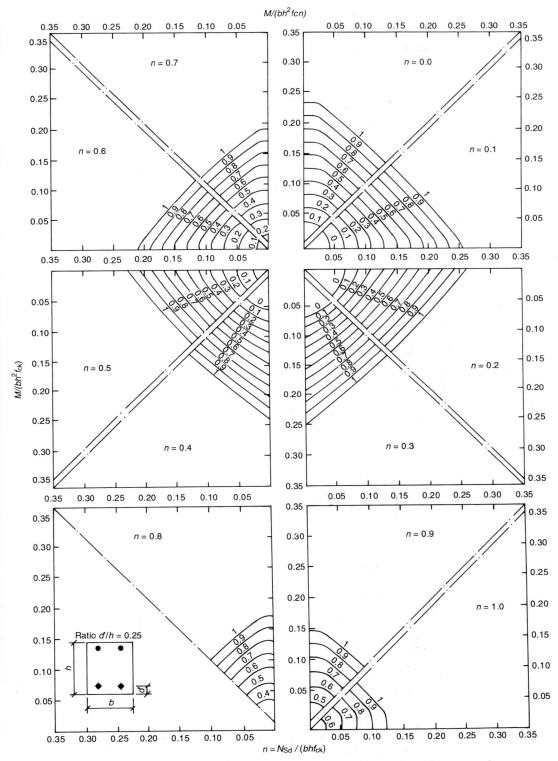

Fig. 5.30. Design chart for biaxial bending of symmetrically reinforced rectangular sections

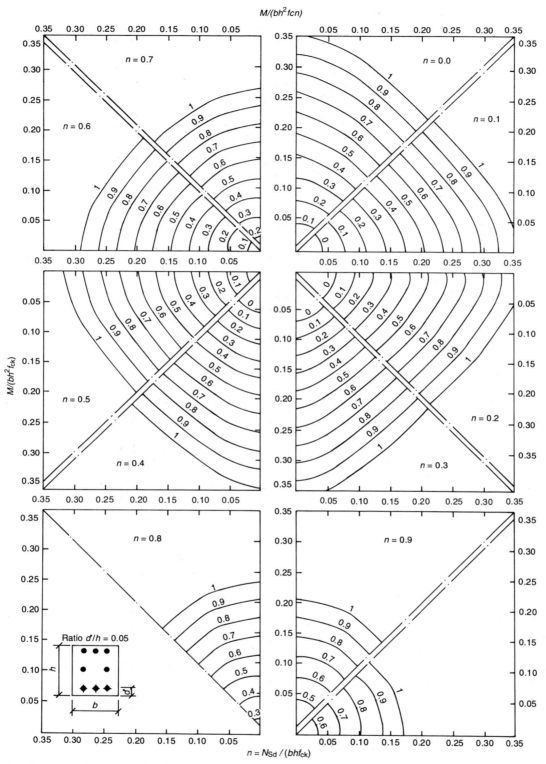

Fig. 5.31. Design chart for biaxial bending of symmetrically reinforced rectangular sections

DESIGN OF SECTIONS FOR BENDING AND AXIAL FORCE

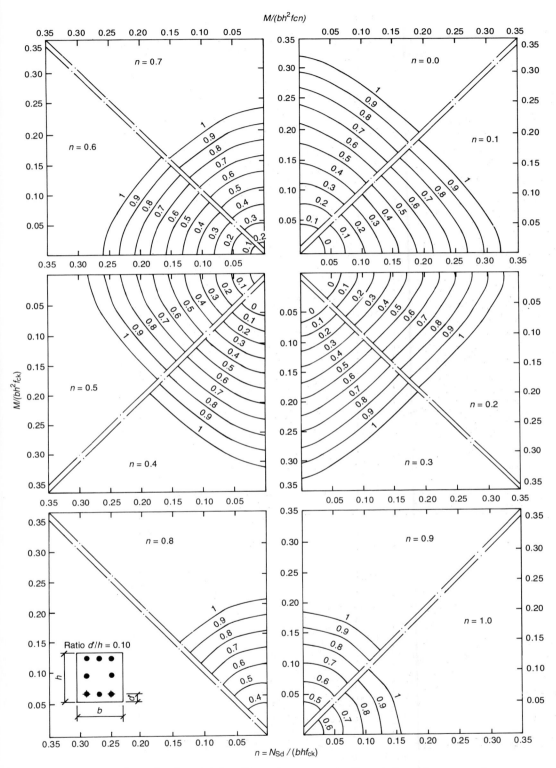

Fig. 5.32. *Design chart for biaxial bending of symmetrically reinforced rectangular sections*

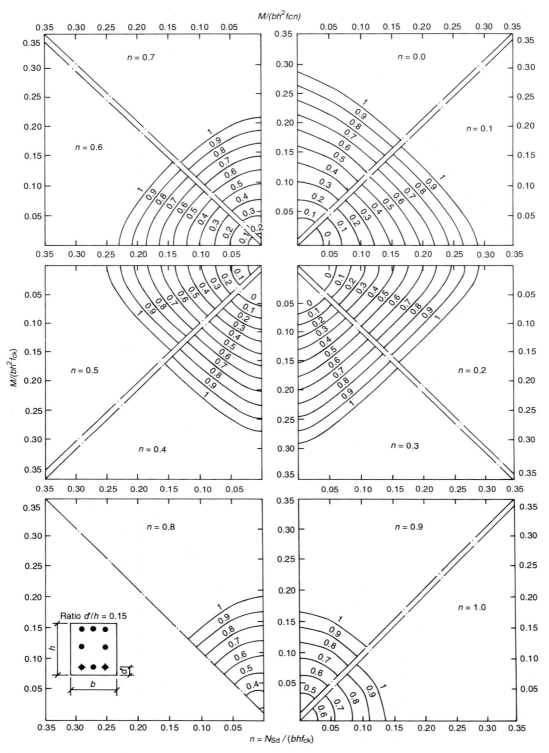

Fig. 5.33. Design chart for biaxial bending of symmetrically reinforced rectangular sections

DESIGN OF SECTIONS FOR BENDING AND AXIAL FORCE

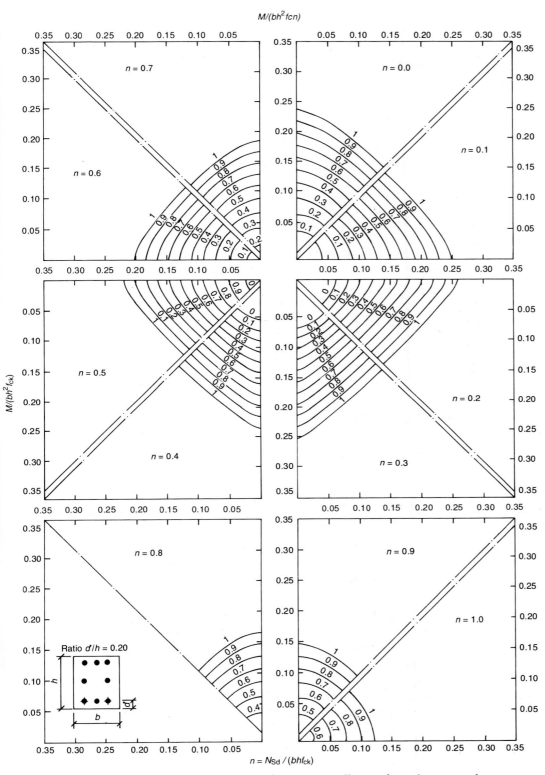

Fig. 5.34. *Design chart for biaxial bending of symmetrically reinforced rectangular sections*

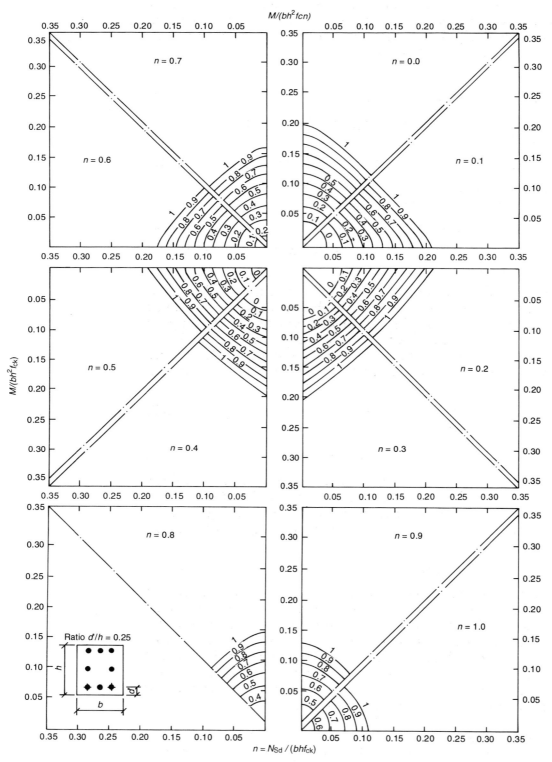

Fig. 5.35. Design chart for biaxial bending of symmetrically reinforced rectangular sections

guess is made of N/N_{uz}, the section is designed, a corrected value of N/N_{uz} is estimated and the process is repeated until a correct solution is obtained. To try to avoid this problem, a simplified method[9] was developed which was shown to give results that were generally close to those obtained by the CP110 approach. The stages in the procedure are as follows

(a) calculate
$$\beta = (b - 1\cdot5d'_b)/(h - 1\cdot5d'_h) \quad (5.24)$$

(b) make a first estimate of M_{ux} from $M_{ux} = M_x + M_y/\beta$
(c) calculate N_{uz} from
$$N_{uz}/N = 1 + M_{ux}/[0\cdot4N(h - 1\cdot5d'_h)] \quad (5.25)$$

(d) hence find a from Table 5.4
(e) calculate a corrected M_{ux} from
$$M_{ux} = [M_x^a + (M_y/\beta)^a]^{1/a} \quad (5.26)$$

(f) design the section as a uniaxially bent section subjected to a moment M_{ux} and an axial load N.

This method has the advantage of being more direct than that in CP110, but is still far from simple. A much simpler but considerably more approximate method has been adopted in BS8110. This is a slightly modified version of a method given by CEB.[10] As in the method described above, design is carried out for an increased uniaxial moment which takes account of the biaxial effects. The required uniaxial moment is obtained from whichever of the following relations is appropriate. If $M_x/h' > M_y/b'$

$$M'_x = M_x + \beta h' M_y/b' \quad (5.27)$$

and if $M_x/h' < M_y/b'$

$$M'_y = M_y + \beta b' M_x/h' \quad (5.28)$$

where M_x and M_y are the design moments about the x and y axes respectively and M'_x and M'_y are the effective uniaxial moments for which the section is actually designed. b' and h' are effective depths as indicated in Fig. 5.23.

The factor β is defined in BS8110 as a function of N/bhf_{cu}. In terms of f_{ck}, it can be obtained from

$$\beta = 1 - N/bhf_{ck} \quad (0\cdot3 < \beta < 1\cdot0) \quad (5.29)$$

This approach has the great advantage of being very simple to apply. Hence, in BS8110, it meant that there was no need to ensure that it was rarely necessary to design for biaxial bending. It is, however, an approximate approach, and it effectively defines the interaction diagram shown in Fig. 5.24. From the code drafting point of view, the problem is whether to adopt a simplified method of biaxial design so that a biaxial design can be easily carried out wherever there is a possibility of biaxial action, or to require a fairly rigorous approach but limit the number of situations where it needs to be applied.

Example 5.4: Biaxially bent column section
A rectangular column section of dimensions 500 mm × 400 mm is to be designed to sustain an axial load of 4800 kN combined with a moment of 245 kN m about the minor axis and 300 kN m about the major axis. The centres of the reinforcing bars may be assumed to be 50 mm from the shorter faces of the section and 60 mm from the longer faces. This gives $h'/h = 0\cdot1$ and $b'/b = 0\cdot15$.

The concrete grade is C30/37 and the reinforcement strength is 500. The reinforcement should be assumed to be fairly uniformly redistributed around the perimeter of the section, and hence the appropriate design charts are Figs 5.15 and 5.16.

The required area of reinforcement will be calculated using both the methods presented above, starting with the more exact method.

The more exact method is iterative, and requires successive estimates to be made of the required reinforcement and the resulting designs to be checked for the biaxial condition using equation (5.23). The first step is to calculate the parameterized moments and axial load. These are

$$N/(bhf_{ck}) = 4800 \times 1000/(500 \times 400 \times 30) = 0.80$$

$$M_x/(bh^2f_{ck}) = 300 \times 10^6/(400 \times 500^2 \times 30) = 0.1$$

$$M_y/(hb^2f_{ck}) = 245 \times 10^6/(500 \times 400^2 \times 30) = 0.1$$

A rough first estimate of the required value of $A_s f_{yk}/(bhf_{ck})$ can be obtained by finding the largest value needed for uniaxial bending. From Fig. 5.16, this is 0.65. Clearly, the true value will be larger than this, say 0.9. For this amount of reinforcement, either Fig. 5.15 or Fig. 5.16 can be used to find $N_{uz}/(bhf_{ck})$ by taking the point where the line corresponding to 0.9 crosses the vertical axis. This value will be found to be 0.135. From this, N/N_{uz} can be calculated as $0.8/1.35 = 0.6$. This can be used to find a value of a from Table 5.4. The appropriate value is 1.66. Figs 5.15 and 5.16 can now be used to obtain values of $M_{ux}/(bh^2f_{ck})$ and $M_{uy}/(hb^2f_{ck})$ corresponding to $N/(bhf_{ck}) = 0.8$ and $A_s f_{yk}/(bhf_{ck}) = 0.9$. These are found to be 0.18 and 0.165 respectively. Equation (5.23) may now be used to check the first estimate

$$(0.1/0.18)^{1.66} + (0.1/0.165)^{1.66} = 0.81$$

This is less than 1.0, so the reinforcement area can be reduced and the procedure repeated. It will be found that the exact solution corresponds to $A_s f_{yk}/(bhf_{ck}) = 0.8$. This gives a reinforcement area of 9600 mm².

The first approximate but more direct method is now tried.

(a) $\beta = (b - 1.5d'_b)/(h - 1.5d'_h) = (400-90)/(500-75) = 0.729$
(b) first estimate of $M_{ux} = M_x + M_y/\beta = 300 + 245/0.729 = 636$
(c) $N_{uz}/N = 1 + M_{ux}/(0.4N(h-1.5d'_h)) = 1 + 636/(0.4 \times 4800 \times 0.425)$
    $= 1.77$
(d) from Table 5.4, $a = 1.60$
(e) corrected $M_{ux} = (300^{1.6} + (245/0.729)^{1.6})^{1/1.6} = 491$
(f) $M_{ux}/(bh^2f_{ck}) = 0.16 N/bhf_{ck} = 0.8$, hence from Fig. 5.16 $A_s f_{yk}/(bhf_{ck}) = 0.8$ and hence $A_s = 9600$ mm².

In this case the answer is identical to the more rigorous calculation. Agreement should generally be close, but the exact agreement in this case is merely coincidental.

Finally, the simplest and most direct method is tried

$$M_x/h' = 300/0.450 = 666 \quad M_y/b' = 245/0.340 = 720$$

Since $M_x/h' < M_y/b'$, equation (5.28) is used. From equation (5.29), $\beta = 1 - 0.8$ but not less than 0.3, so $\beta = 0.3$. Hence

$$M'_y = 245 + 0.3 \times 300 \times 340/450 = 313 \text{ kN m}$$

$$M_y/(hb^2f_{ck}) = 0.13, \quad N/(bhf_{ck}) = 0.8$$

Hence, from Fig. 5.16 $A_s f_{yk}/(bhf_{ck}) = 0.77$ and hence $A_s = 9240$ mm². This method has thus produced an answer within 4% of those given by the more rigorous methods.

5.2.6. Design of prestressed sections
The design of prestressed sections introduces no major new problems except that allowance must be made for the prestrain in the prestressing steel. This is effectively accomplished by fixing a false origin for the stress–strain curve for the prestressing steel at a point corresponding to zero stress and a strain equal to the prestrain, as shown in Fig. 5.25.

5.2.7. Brittle failure and hyperstrength
Clause 4.3.1.3 states that brittle failure shall be avoided and that flexural strengths above those calculated on the basis of the assumptions given in *clause 4.3.1* shall not be used in design, even where tests show such strengths to occur.

Clause 4.3.1.3
Clause 4.3.1

Brittle failure on the formation of the first crack may occur in very lightly reinforced members. It will happen when the strength of the section based on the tensile strength of the concrete is greater than the strength of a cracked section where the strength is defined by the yield force in the reinforcement. If this occurs, there is a sudden drop in load-carrying capacity on occurrence of the first crack. It is not difficult to formulate equations defining this condition. For pure bending, the equation defining the condition under which brittle failure will occur is

$$f_t Z > A_s f_y d \qquad (5.27)$$

where f_t is the tensile strength of the concrete, Z is the section modulus ($= bh^2/6$ for a rectangular section), A_s is the area of tension reinforcement, f_y is the yield strength of the reinforcement and d is the effective depth.

Equation (5.27) is a slight approximation since it assumes that the lever arm of the internal forces is equal to the effective depth d, whereas in fact it will be slightly less than this.

For a rectangular section, the condition for avoiding brittle failure can be re-expressed as

$$A_s/bd > (1/6)(f_t/f_y)(h/d)^2 \qquad (5.28)$$

From this it will be seen that brittle failure can be avoided by defining an appropriate minimum reinforcement ratio. Minima are given in *clause 4.4.2 and Chapter 5* which are sufficient to avoid brittle failure. In fact, the provisions of *clause 4.4.2.2* are formulated in the same way as outlined above. These provisions are discussed in more detail in a later chapter. Thus, provided the rules in various parts of EC2 for minimum steel ratios are satisfied, no further action is required of the designer to avoid brittle failure.

Clause 4.4.2
Chapter 5
Clause 4.4.2.2

Hyperstrength is a reflection of the same phenomenon. The strength of a section calculated using the assumptions given in *clause 4.3.1* takes account only of the concrete in compression and the reinforcement in tension. It follows that, if the conditions in the section are such that brittle failure will occur, the actual strength of the section will be greater than the calculated strength since the actual strength depends on the tensile strength of the concrete. Since brittle failure is not permitted, it follows that this extra strength cannot be used in design even though a test would show the extra strength to exist. No action is required of the designer to avoid hyperstrength beyond the measures necessary to avoid brittle failure.

Clause 4.3.1

CHAPTER 6

Shear, punching shear and torsion

6.1. Shear: general
Shear is dealt with in *clause 4.3.2*. Punching shear is dealt with separately in *clause 4.3.4*.

Clause 4.3.2
Clause 4.3.4

More research has probably been carried out into shear behaviour than into any other mode of failure of reinforced or prestressed concrete. However, considerable areas of uncertainty and disagreement remain. Furthermore, unlike flexural behaviour, there is no generally accepted overall model describing shear behaviour. Despite these uncertainties, design for shear can be carried out with confidence for all normal members because the design methods given in codes have had the advantage of being tested against, and adjusted to fit, a very large body of experimental data. The first part of this chapter outlines shear behaviour as it is currently understood, with an indication of how the provisions in EC2 reflect this.

6.2. Background to the EC2 provisions
Four aspects of the strength of sections subjected to shear loading are considered here in turn:

- strength of members without shear reinforcement
- strength of members with shear reinforcement
- maximum shear that can be carried by a member
- behaviour close to supports.

6.2.1. Members without shear reinforcement
This area is probably of limited importance for beams, where some shear reinforcement will always be provided, but it is of major importance for slabs, where it is often very inconvenient to provide shear reinforcement. It is particularly important for design for punching (see section 6.4). This has been the most researched area of shear performance, but no generally accepted theory describing the ultimate behaviour of a member without shear reinforcement has yet been developed. The formulae given in codes should therefore be considered to be basically empirical. Because of the amount of testing that has been carried out, the effect of the major variables can be clearly established and the resulting formulae can be considered as highly reliable for normal types of element.

The principal variables governing the shear strength of members without shear reinforcement are the concrete strength, the reinforcement ratio and the depth of the member.

It is fairly self-evident that concrete strength will have an influence on shear strength, although the nature of the failure suggests that the tensile strength rather than the compressive strength is likely to be important. EC2 predicts that the shear strength will be directly proportional to the tensile strength of the concrete. The relationship of tensile strength and compressive strength assumed in EC2 is that tensile strength is proportional to the compressive strength to the power of 2/3. Other codes, such as BS8110 and the ACI code, predict a lesser influence. The ACI code predicts the shear strength as varying in proportion to the square root of the compressive strength; BS8110 gives a strength proportional to the cube root of the compressive strength combined with a cut-off at a cube strength of 40 N/mm². It may well be that EC2 should not be extrapolated to concrete strengths above C50/60, the maximum it currently covers, without a very careful appraisal of the influence of higher strengths.

Shear strength increases with increasing reinforcement ratio, but the rate of increase reduces as reinforcement ratio increases. This behaviour can be modelled in various ways. EC2 uses a bilinear relationship, with the strength proportional to $(1 \cdot 2 + 40\rho)$ up to a maximum value of $0 \cdot 02$. BS8110 employs a cube root relation of shear strength and steel percentage up to a maximum ρ value of $0 \cdot 03$. The experimental data are so scattered that it is not possible to differentiate practically between these two methods.

Absolute section depth is found to have a significant influence on shear strength over and above the influence expected from normal geometrical scaling (i.e. there is a size effect). This effect is sufficiently large to be worth taking into account in the design of shallow members such as slabs. Most recent codes of practice therefore have a term in their equations to allow for this, which gives a higher shear strength for shallow members.

Taking these factors into account, EC2 gives the following equation for the strength of sections without shear reinforcement

$$V_{Rd1} = b_w d(\tau_{Rd} k (1 \cdot 2 + 40\rho_1) + 0 \cdot 15 \sigma_{cp}) \tag{6.1}$$

where

$$\tau_{Rd} = 0 \cdot 25 f_{ctk0 \cdot 05}/\gamma_c \tag{6.2}$$

$k = 1$ for members where more than 50% of the bottom steel has been curtailed, otherwise $k = (1 \cdot 6 - d)$ where d is in metres; $\rho_1 = A_{s1}/b_w d \leq 0 \cdot 02$, and σ_{cp} = average longitudinal stress.

6.2.2. Members with shear reinforcement

There is a generally accepted model for the prediction of the effects of shear reinforcement. This is the 'truss model' illustrated in Fig. 6.1 for the commonest case where vertical links are used. In this model the top and bottom compression and tension members are the concrete in the compression zone and the tension steel respectively. The members connecting the top and bottom members are represented by steel tension members and virtual concrete 'struts'. The truss in Fig. 6.1 can be analysed to give the following forces in the various members

$$F_1 = N/2 + V(a_v/z - 0 \cdot 5 \cot \theta) \tag{6.3}$$

$$F_3 = N/2 - V(a_v/z + 0 \cdot 5 \cot \theta) \tag{6.4}$$

$$F_2 = V/\sin \theta \tag{6.5}$$

$$F_4 = V \tag{6.6}$$

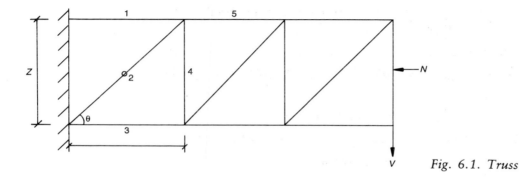

Fig. 6.1. Truss

Two facts can be observed from this analysis. First, the forces in the vertical tension member (4, Fig. 6.1) and the compression strut are both independent of the axial force N. Second, the forces in the compression and tension chords both differ from the force calculated from the moment alone by $0.5V \cot \theta$. ($Va_v = M$ and the forces in the upper and lower chords are given by bending theory as $\pm M/z$.)

In developing this truss into a design approach for a reinforced or prestressed concrete beam, it is necessary to consider it as a 'smeared' truss as illustrated in Fig. 6.2. In this idealized truss, the vertical stirrups are represented by a uniform vertical tensile stress of $\rho_w f_{yk}$ or a vertical force of $\rho_w f_{yk} b_w$ per unit length of the beam. This assumes that, at failure, the stirrups yield. The virtual compressive strut is replaced by a uniform uniaxial compressive stress of σ_c acting parallel to the line of action of the strut over the concrete between the centroid of the tension reinforcement and the centre of compression. A complete analysis of this smeared truss may be carried out by considering the equilibrium of two sections.

(a) Vertical equilibrium across a section parallel to the virtual compression strut. On this section, all the shear is supported by the shear reinforcement (section A–A in Fig. 6.2). This gives

$$V = \rho_w f_{yk} b_w z \cot \theta \tag{6.7}$$

(b) Vertical equilibrium across a section perpendicular to the strut (section

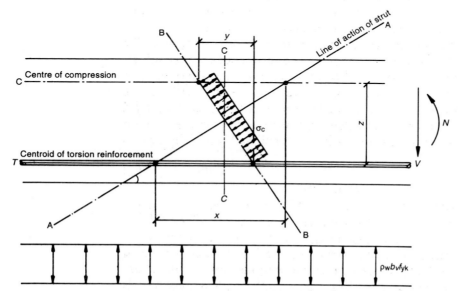

Fig. 6.2. Idealization of a beam with shear reinforcement as a smeared truss

B—B in Fig. 6.2). Across this section, some of the shear is carried by the strut and some by the stirrups.

$$\text{total strut force} = F_c = \sigma_c b_w z / \cos\theta \tag{6.8}$$

The horizontal projection of the section is given by $z \tan\theta$, hence equilibrium is given by

$$V = \rho_w f_{yk} b_w z \tan\beta + F_c \cos\theta$$

Since, from equation (6.7), $\rho_w f_{yk} b_w z = V \tan\theta$, this can be rewritten as

$$V = V \tan^2\theta + \sigma_c b_w z \tan\theta$$

which can be rearranged to give

$$V = \frac{\sigma_c b_w z}{(\cot\theta + \tan\theta)} \tag{6.9}$$

These equations can be converted to design values by simply replacing the characteristic material properties by the appropriate design values. σ_c is replaced by νf_{cd}: the parameter ν is an efficiency factor which allows for the actual distribution of the stress within the strut. Equation (6.9) then gives the maximum shear that can be carried by a section before failure by crushing of the notional compression struts.

Equations for inclined shear reinforcement can be derived in a similar way by replacing the vertical tension by a uniform tension inclined at an angle to the horizontal.

The design forms of equations (6.7) and (6.9) appear in EC2 as *equations (4.26) and (4.27)*.

There are two areas where methods in various codes differ. The first is in the choice of a value for the truss angle θ. The second is whether the truss should carry all the shear force or part of the shear can be considered to be carried by the concrete. EC2 permits the use of two methods: the 'standard method', which uses a fixed truss angle of $45°$ and assumes that the shear reinforcement is required only to carry the excess shear force above V_{Rd1}, and the 'variable truss angle method'. In the latter, all the shear is carried by the shear reinforcement but the truss angle θ can take any value between $\cot^{-1}0.4$ and $\cot^{-1}2.5$. The variable strut angle approach is considered to be the more rigorous of the two methods, and also the more economical. However, as expressed in EC2, it is open to a misunderstanding. EC2 implies that the designer may select any strut angle he chooses between the specified limits. This concept of free choice does not, however, reflect the behaviour of a beam. Beams will fail in a manner corresponding to a strut angle of roughly $\cot^{-1}2.5$ unless constrained by the detailing or the geometry of the system to fail at some steeper angle. A steeper angled failure could be induced by the way in which the tension steel was curtailed, or where the load is so close to the support that only a steeper failure can occur. Fig. 6.3 illustrates this aspect of behaviour.

6.2.3. Maximum shear strength of a section

Section 6.2.2 is concerned with the behaviour of the tension members in the truss. The capacity of these can be increased to whatever level is required simply by increasing the amount of reinforcement provided. However, the capacity of the compression members (the virtual diagonal struts) cannot be so easily varied. It is the strength of these struts that provides an absolute upper limit to the shear that can be supported by the beam. The shear supported by these members is derived above as

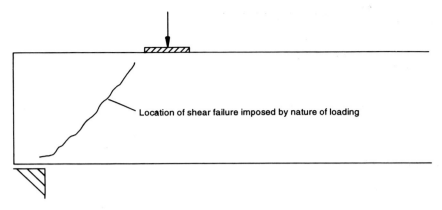

Fig. 6.3. Shear failure close to a support

$$V = \sigma_c b_w z / (\cot \theta + \tan \theta)$$

At the ultimate load, the strut will crush and the average stress in the strut σ_c can be expected to be proportional to the compressive strength of the concrete. This is achieved by defining the limiting value of σ_c as νf_{cd}, where ν is an empirically obtained efficiency factor which can be considered to take account of the actual distribution of stress across the section at ultimate. Thus

$$V_{Rd2} = b_w z \nu f_{cd} / (\cot \theta + \tan \theta) \qquad (6.11)$$

where V_{Rd2} is the maximum shear capacity of the section.

Where the 'standard method' is used, $\cot \theta = \tan \theta = 1 \cdot 0$. Introducing the further simplification that $z = 0 \cdot 9d$ gives

$$V_{Rd2} = 0 \cdot 45 \nu f_{cd} b_w d \qquad (6.12)$$

Both these equations are appropriate for vertical shear reinforcement.

The lower-bound theory of plasticity states that, provided there is adequate ductility, the internal forces within a section will adjust themselves to accommodate the maximum possible load. From equation (6.7), the smaller the angle θ, the greater is the shear capacity based on the shear reinforcement. However, the shear capacity based on the crushing strength of the strut, given by equation (6.11), decreases with decreasing values of θ below $45°$. Hence the maximum capacity corresponds to the situation where the capacity based on the shear reinforcement just equals the capacity based on the strength of the strut. This implies that the actual conditions at failure can be established by using equation (6.12) to estimate the value of θ for which $V = V_{Rd2}$ and then using this value of θ to obtain the required amount of shear reinforcement. In EC2, this method is adopted, but only if the calculated value of θ is greater than $\cot^{-1} 2 \cdot 5$. Note that V_{Rd2} reaches a maximum value when $\cot \theta = 1$ and hence, again from the lower-bound theory of plasticity, values of θ greater than $45°$ will occur only if other factors constrain the failure to occur at such an angle.

6.2.4. Shear capacity enhancement near supports

There is very extensive test work that shows that much higher shear strengths than are given by equation (6.1) can be obtained in short members such as corbels or in beams where the load is applied close to the support. The reason for this is simply that, for any sections closer to the support than the critical section defined by the 'natural' strut angle of $\cot^{-1} 2 \cdot 5$ (see Fig. 6.4), a substantial proportion of the load will be carried through to the support directly by the strut and not

SHEAR, PUNCHING SHEAR AND TORSION

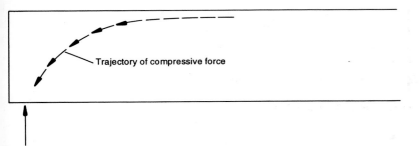

Fig. 6.4. Direct transfer of load to support

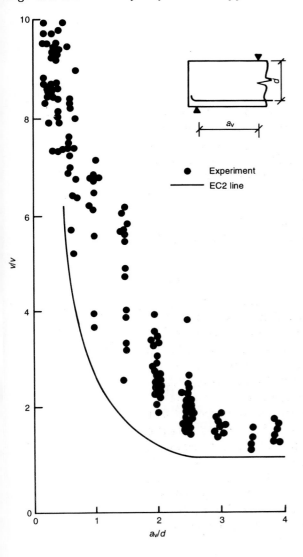

Fig. 6.5. Experimental results for shear strength near supports

by way of the normal actions of shear and bending. The closer the load to the support, the greater is the proportion of the load that will be transmitted to the support in this way. Experimental evidence suggests that a convenient method of treating this problem is to assume that the shear carried by this mechanism is a function of d/x, where x is the distance from the face of the support to the face of the load (see Fig. 6.4). Fig. 6.5 shows experimental results for failures close to supports and the form of the interaction used by EC2, which deals with this problem by increasing the design shear resistance of the concrete section V_{Rd1} by a factor $\beta = 2 \cdot 5d/x$.

A further point to note is that, clearly, this enhancement can be applied only where the load is applied to the top face of the beam and the support is at the bottom.

6.2.5. Summary
It has been attempted above to give a brief picture of the phenomenon of shear failure and the factors that influence it. This should help designers to understand the provisions of EC2 and apply them with confidence. It should be understood, however, that this picture is simplified: shear is still not fully understood, and remains a very complex phenomenon.

6.3. Summary of the provisions in *clause 4.3.2*

Clause 4.3.2

The flow chart in Fig. 6.6 gives an overview of the provisions in this chapter, and should act as a guide in their application. Figs 6.7–6.9 are charts which will enable the basic quantities used in the shear calculation to be read off for most normal situations that will be encountered for reinforced concrete.

Figure 6.7 can be used to obtain values of $V_{Rd1}/b_w d$ for various values of longitudinal reinforcement ratio and concrete grade. These values will need to be multiplied by the factor k for elements with effective depths below 600 mm. Values for k are given in Fig. 6.8. For members supporting an axial force (i.e. a column or a prestressed beam), the values of V_{Rd1} obtained from Figs 6.7 and 6.8 are increased by addition of a shear equal to $0 \cdot 15 N/A_c b_w d$. In this expression, N is the axial force and A_c is the cross-sectional area of the element.

Where shear reinforcement in the form of vertical stirrups is to be provided, Fig. 6.9 may be used to calculate the required amount. Fig. 6.9 permits the calculation of vertical shear reinforcement using the variable strut inclination method given in *clause 4.3.2.4.4*. The lines are derived from *equations (4.26), (4.27)* and *(4.21)* using the procedure set out in *clause 4.3.2.4.4(4)*. To use the chart, first estimate the value of $V_{sd}/b_w d$. The design relationship is then defined by two curves: (*a*) a straight line defined by the maximum permissible value of cot θ, and (*b*) a curve defined by the grade of concrete. The maximum value that may be taken for cot θ is $2 \cdot 5$ where the longitudinal reinforcement is constant and $2 \cdot 0$ where it has been curtailed. The distance over which the reinforcement must be constant for a value of $2 \cdot 5$ to be acceptable is not stated, but it cannot mean the whole beam, since a curtailment at the opposite end of a beam to the end being considered can hardly affect the shear strength. It is suggested that, provided there is no curtailment of the tension steel within $2 \cdot 5d + l_{b.net}$ of the section considered, cot θ may be taken as $2 \cdot 5$. The broken line defining the end of the curves gives the maximum amount of vertical shear reinforcement that may be used. It corresponds to cot $\theta = 1$ or $A_{sw} f_{ywd}/b_w s = 0 \cdot 5 f_{cd}$. The two conditions are equivalent and correspond to the maximum possible value of V_{Rd2} given by *equation (4.26)*. Minimum areas of shear reinforcement are not marked in Fig. 6.9, as this would overcomplicate the drawing, but values can be obtained from *Table 5.5*. These values are (for concrete grade C20/25 and below)

Clause 4.3.2.4.4
Clause 4.3.2.4.4(4)

$$A_{sw} f_{ywd}/b_w s \geq 0 \cdot 35$$

(for concrete grades C25/30–C35/45 inclusive)

$$A_{sw} f_{ywd}/b_w s \geq 0 \cdot 52$$

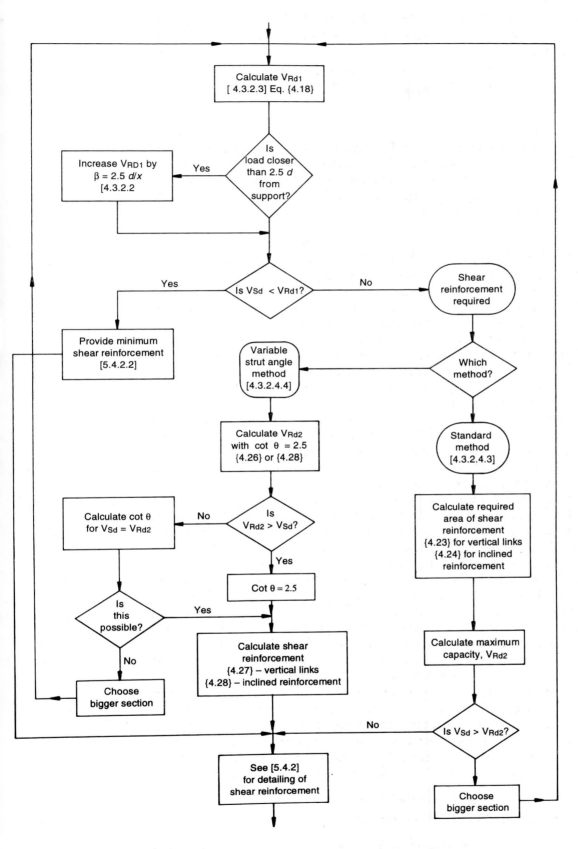

Fig. 6.6. *Flow chart for shear design*

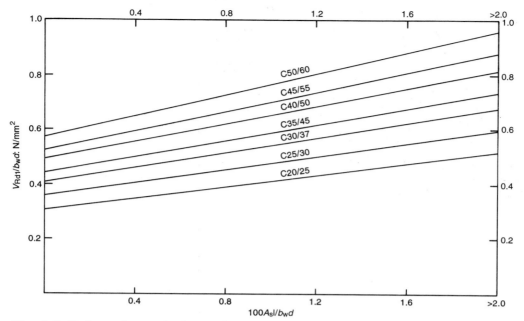

Fig. 6.7. Values of $V_{Rd1}/b_w d$ as a function of reinforcement ratio and concrete strength

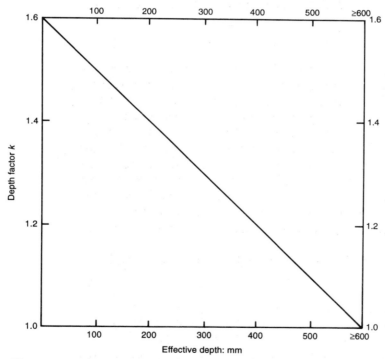

Fig. 6.8. Factor to correct V_{Rd1} for the effects of section depth

(for concrete grades C40/50 and above)

$$A_{sw} f_{ywd}/b_w s \geq 0.64$$

In selecting the critical section for shear design, it should be noted that *clause 4.3.2.2(10)* indicates that V_{sd} may be evaluated at a distance d from the face of

Clause 4.3.2.2(10)

SHEAR, PUNCHING SHEAR AND TORSION

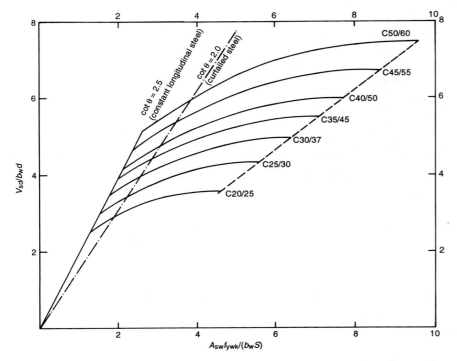

Fig. 6.9. Vertical shear reinforcement calculated using the variable strut inclination method

the support for beams with uniformly distributed loading. This takes account of the increased shear resistance near the supports. For concentrated loads, this effect is accounted for by the shear force reduction factor given by *equation (4.17)*.

Again, it should be noted that all the equations derived and presented in this chapter deal only with the case of vertical shear reinforcement. No new principles are involved in the extension of the derivations to the case of inclined shear reinforcement; it simply complicates the equations and makes the derivation less clear. Furthermore, from the practical point of view, the use of inclined shear reinforcement is becoming increasingly rare due to the considerably increased expense of the reinforcement fixing and difficulties in maintaining tolerances. Should inclined shear reinforcement be needed, the necessary equations are all presented in EC2 itself.

The best way to illustrate the design procedure is by means of examples.

Example 6.1: Rectangular section

The simply supported beam illustrated in Fig. 6.10 is to be designed for shear using both the 'standard' method and the 'variable strut angle' method. The design for flexure has resulted in a requirement for 2200 mm² of longitudinal reinforcement. This is provided by two 32 mm and two 20 mm bars, giving an area of 2236 mm².

The shear force at the supports is given by

$$(1 \cdot 35 \times 40 + 1 \cdot 5 \times 20) \times 5/2 = 210 \text{ kN}$$

However, the design shear may be taken as the shear a distance d from the face of the support. Assuming that the distance from the support centre line to the face of the support is 75 mm, the critical design shear is that at a distance 75 + 325 = 400 mm from the support centre line. This shear can be calculated as

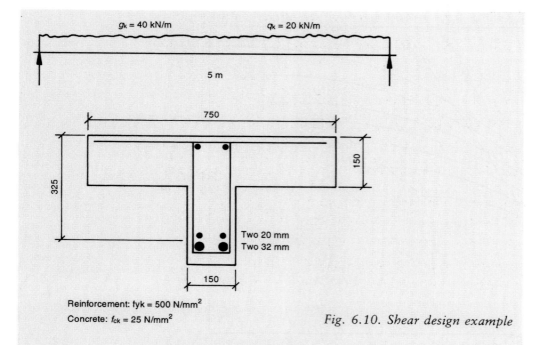

Fig. 6.10. Shear design example

$$V_{Sd} = 210\,(5000/2 - 400)/(5000/2) = 176.4 \text{ kN}$$

Now calculate V_{Rd1}. The reinforcement ratio $A_{s1}/b_w d = 2236/150/325 = 4.59\%$. From Fig. 6.7, the basic value of $V_{Rd1}/b_w d$ is 0.6 N/mm², and from Fig. 6.8, the correction factor for depth is 1.275. Hence

$$V_{Rd1} = 0.6 \times 1.275 \times 150 \times 325 = 37.3 \text{ kN}$$

This is much less than V_{Sd}, hence designed shear reinforcement is needed. From here on, the two design methods differ.

Standard method
The contribution required from the shear reinforcement is given by

$$V_{wd} = V_{Sd} - V_{Rd1} = 176.4 - 37.3 = 139.1 \text{ kN}$$

The required amount of shear reinforcement can now be calculated from *equation (4.23)*

$$A_{sw}/s = V_{wd}/(0.9 d f_{ywd}) = 139.1 \times 1000/500 \times 1.15/325/0.9 = 1.094$$

This could be provided by 10 mm dia. stirrups at 130 mm centres.
 From equation (4.25), V_{Rd2} can be calculated as

$$V_{Rd2} = 0.5 \times 0.575 \times 25/1.5 \times 150 \times 0.9 \times 325 = 210.23 \text{ kN}$$

This is greater than V_{Sd}, hence the design is satisfactory. However, since $V_{Sd} > \frac{2}{3} V_{Rd2}$, *equation (5.19)* imposes a maximum spacing on the stirrups of $0.3d = 97.5$ mm. This suggests that smaller diameter stirrups than the 10 mm stirrups assumed above would be better. 8 mm dia. stirrups at 90 mm centres will be satisfactory. *Clause 5.4.2.2*

The centre section of the beam will require only the minimum reinforcement specified in *Table 5.5*. From *Table 5.5*, the minimum value of A_{sw}/s is *Clause 5.4.2.2*

$$0.0011 \times 150 = 0.165$$

From *equations (4.22) and (4.24)*, this corresponds to a design shear force of

$V_{Rd3} = 37.3 + 0.165 \times 0.9 \times 325 \times 500/1.15/1000 = 58.3$ kN

The design shear force will be less than this over the central 1.4 m of the beam.

Variable strut angle method

$$V_{sd}/b_w d = 176.4 \times 1000/150/325 = 3.61$$

Fig. 6.9 gives the required value of $A_{sw}f_{ywk}/b_w s$ as 2.6. Hence

$$A_{sw}/s = 2.6 \times 150/500 = 0.78$$

The most economical solution that will not violate the maximum stirrup spacing rules would be to use 6 mm dia. stirrups at 70 mm centres. This will be substantially more economical than the standard method.

The capacity of minimum stirrups can be calculated from *equation (4.27)* as

$$V_{Rd3} = 0.165 \times 0.9 \times 325 \times 500/1.15 \times 2.5 = 52.4 \text{ kN}$$

This indicates that minimum stirrups will be adequate over the central 1.25 m of the span. This is a slightly smaller length than can be achieved using the standard method.

6.4. Punching shear

6.4.1. General

Punching shear is a local shear failure around a concentrated load on a slab. The most common situation where punching shear has to be considered is the region immediately surrounding a column in a flat slab. Design for punching shear is covered in *clause 4.3.4*.

Clause 4.3.4

Punching shear failures may be considered to be shear failures rotated around the loaded area so as to give a failure surface that has the form of a truncated cone. This is illustrated in Fig. 6.11. The 'critical section' for shear failure in a beam is transformed into a 'critical perimeter' when punching shear is considered. This conversion of the problem from basically two-dimensional to three-dimensional does not change the basic phenomenon as described in section 6.3 above, although a number of practical points need further consideration. The first point to be considered is the definition of the critical perimeter.

6.4.2. Critical perimeter

The starting point for design for punching shear is the definition of the critical perimeter. This is more important than the selection of the critical section in a beam because, as perimeters closer to the loaded area are considered, the length of the perimeter rapidly gets shorter and hence the shear force per unit length of the perimeter rapidly increases. Observation of failures shows that the outer perimeter of a punching failure takes the general form shown in Fig. 6.12(a). For this reason, EC2 proposes the idealized form shown in Fig. 6.12(b). Some national codes (e.g. the UK Code, BS8110) employ a rectangular perimeter. This is less realistic but slightly easier to use. Another advantage of the rectangular perimeter is that when reinforcement is needed, it must be provided within the perimeter. It is much easier to do this with a rectangular perimeter, since reinforcement is usually provided on a rectangular grid. In fact, as will be seen below, since the design equations for punching shear are basically empirical, the precise choice of the shape of the perimeter is unimportant.

Once a basic form for the perimeter has been chosen, it is necessary to consider the distance from the loaded area at which it should be located. Since the treatment

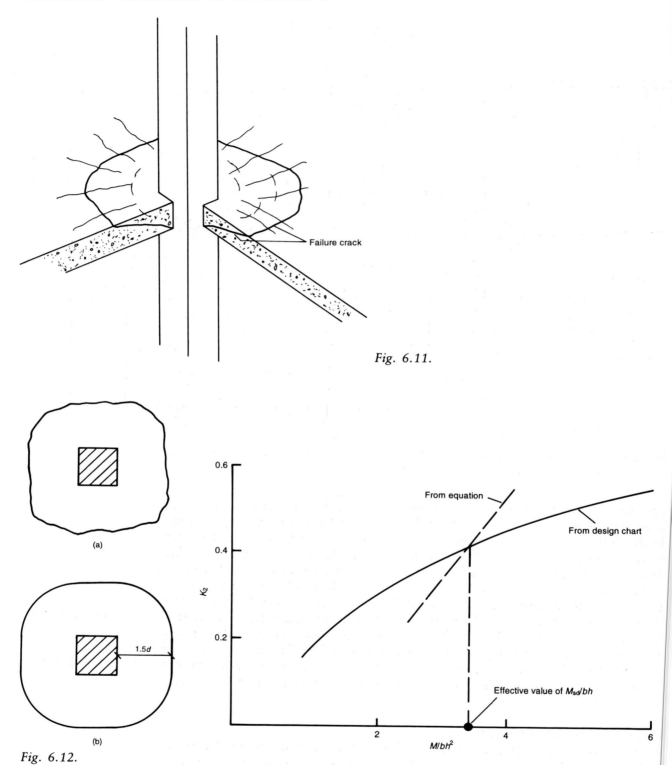

Fig. 6.11.

Fig. 6.12.

of shear is empirical, there is no absolutely defined answer to this question. The same strength could be obtained by using a high value for the shear strength V_{Rd1} combined with a short perimeter close to a column or by using a lower shear strength combined with a longer perimeter further from the column. Different codes have adopted different solutions to this question: the UK code takes a

SHEAR, PUNCHING SHEAR AND TORSION

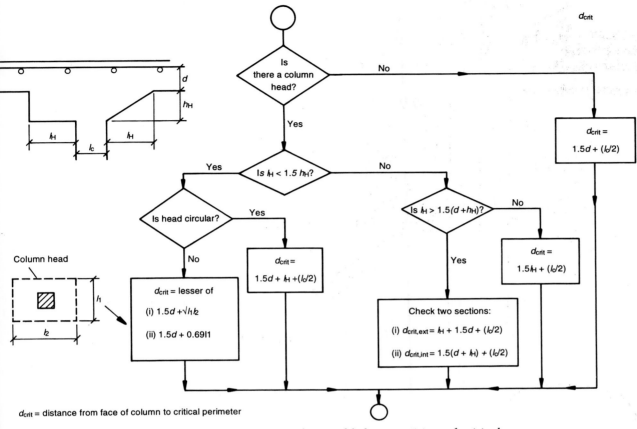

Fig. 6.13. *Interpretation of clause 4.3.4.4 in EC2 for establishing position of critical perimeter where column heads are used*

perimeter $1 \cdot 5d$ from the face of the column or loaded area; the US ACI code uses a perimeter $0 \cdot 5d$ from the face of the column but with a much higher stress. In the drafting of EC2, it was decided that the shear strength should be given by the same formula for punching shear as is used for shear in beams. Having decided on this and on the shape of the perimeter, the distance of the perimeter from the loaded area can be established from test data. This led to a value of $1 \cdot 5d$. The interpretation of this to define the actual perimeter is dealt with very thoroughly in *clause 4.3.4.2.2*, and needs no further discussion here. A question that does require some further clarification is the definition of the critical perimeter in cases where there is a 'drop' or 'column head' around the column. This is covered in *clause 4.3.4.4* and *Figs 4.22 and 4.23*, but is not easy to follow there. The flow chart in Fig. 6.13 is intended to clarify the interpretation of these provisions.

Clause 4.3.4.2.2

Clause 4.3.4.4

6.4.3. Design shear force

It is usual to assume in design that the distribution of shear force around the critical perimeter is uniform. In fact this is not the case, particularly for a slab–column connection where there is a moment transfer between the slab and the column. In such cases, a rigorous analysis would show that the distribution of shear varied markedly around the perimeter and was accompanied by torsional moments. Extensive experimental work shows that punching shear strength can be significantly reduced where substantial moment transfer occurs. A way of dealing with this in design is to increase the design shear force by a factor that is a function of the geometry of the critical perimeter and the moment transferred. The UK Code, BS8110, includes formulae for enhancing the design shear in this way. The provisions, which are basically empirical, are as follows.

(a) For internal columns, design is carried out for an effective design shear where

$$V_{\text{eff}} = V_{\text{Sd}}(1 + 1 \cdot 5 M_t / V_{\text{Sd}} X)$$

in which V_{eff} is the effective shear force allowing for the effects of moment transfer, V_{Sd} is the design shear force, M_t is the moment transferred between slab and column, and X is the overall breadth of slab within the critical perimeter parallel to the axis of bending (see Fig. 6.14).

(b) For edge columns where bending is about an axis perpendicular to the edge, the effective shear is given by

$$V_{\text{eff}} = V_{\text{Sd}}(1 \cdot 25 + 1 \cdot 5 M_t / V_{\text{Sd}} X)$$

(c) For all other situations

$$V_{\text{eff}} = 1 \cdot 25 V_{\text{Sd}}$$

EC2 recognizes this problem but adopts a simplified approach, simply proposing an increase of 15% for the shear at internal columns, 40% for edge columns and 50% for corner columns. *Clause 4.3.4.3(4)* states, however, that other values may be used if based on a more rigorous analysis. On the basis of this provision, it would seem reasonable, where the moment transfer has been calculated, to use the expressions from BS8110 if these give a more economical result.

Clause 4.3.4.3(4)

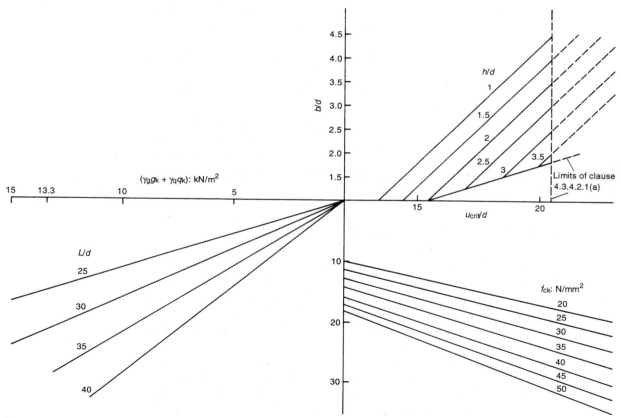

Fig. 6.14. *Design chart to assess load capacity of slab without punching shear reinforcement*

6.4.4. Shear resistance of slabs without shear reinforcement

The basic relationship of shear resistance and reinforcement ratio, concrete strength and slab depth is the same as that used for beams, but with the limitation that the reinforcement ratio should not be taken as greater than 0·15. Because of the two-way nature of the slab, it is necessary to introduce rules for the assessment of the average effective depth and average reinforcement ratio. An additional difference from beams is that, for punching shear, the maximum shear that can be supported by a slab V_{Rd2} is taken as $1·6 V_{Rd1}$. This is much more conservative than the limit for beams. Provided that the shear is less than V_{Rd1}, no shear reinforcement is needed.

An attempt has been made to produce a design chart to establish whether or not shear reinforcement is needed around internal columns (Fig. 6.14). This is not a rigorous representation of the provisions of EC2, but has been developed from a parameter study run using these provisions. The mode of operation of the chart is as follows.

(a) Calculate the ratio of the smaller and larger dimensions of the column to the effective depth of the slab, b/d and h/d respectively.
(b) Enter the chart at the appropriate value of b/d and read across to the appropriate line for h/d, then drop down through the horizontal axis to the line corresponding to the required concrete strength. If desired, the length of the critical perimeter can be read off the horizontal axis.
(c) Read across to the line in the third quadrant corresponding to the span/effective depth ratio, and then upwards to the horizontal axis to read off the design load that the slab can withstand without shear reinforcement.

In the first quadrant, lines are drawn corresponding to the limits of validity of the method given in *clause 4.3.4.2.1(a)*. These limits are as follows.

Clause 4.3.4.2.1(a)

(a) The larger dimension of the column should not exceed twice the smaller dimension. This limit is given by the solid sloping line.
(b) The perimeter of the loaded area should not exceed $11d$. This limit is given by the broken vertical line.

The second limit should possibly not be considered as absolute, since EC2 implicitly ignores it in at least one situation. This is where a drop is used where EC2 requires a check to be carried out on a perimeter $1·5d$ outside the drop. This effectively treats the drop as the loaded area. It is most unlikely that the perimeter of a drop would be less than $11d$.

6.4.5. Reinforcement for punching shear

Shear reinforcement is provided using the standard method as given in *clause 4.3.2.4.3*, except that the lever arm is effectively assumed to be d rather than $0·9d$. The calculated area of shear reinforcement should be provided within the critical perimeter. Where vertical reinforcement is used, the first set of stirrups should be not more than $0·5d$ from the face of the column or loaded area. Subsequent perimeters of stirrups should be located not more than $0·75d$ from the first. The detailing rules for shear reinforcement are given in *clause 5.4.3.3*.

Clause 4.3.2.4.3

Clause 5.4.3.3

Once it has been found that shear reinforcement is necessary on the perimeter located $1·5d$ from the face of the loaded area, further checks should be carried out on successive perimeters located at intervals of $0·75d$ outside this first perimeter until a perimeter is reached that does not require reinforcement. Where reinforcement is required for perimeters further than $1·5d$ from the face of the loaded area, the reinforcement should be located between the perimeter considered

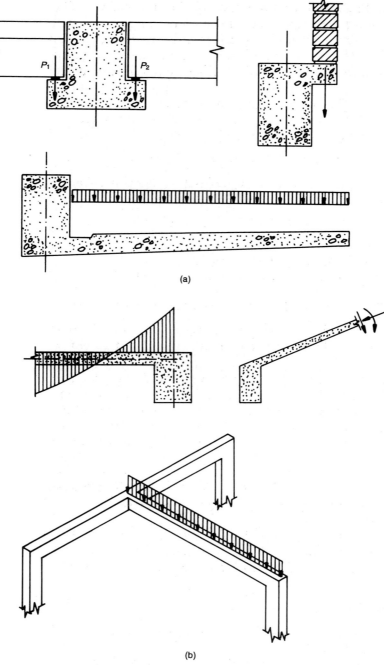

Fig. 6.15. (a) Examples of primary or equilibrium torsion; (b) torsion in statically indeterminate structures

and a perimeter $1 \cdot 5d$ inside it. Reinforcement detailed to resist shear on other perimeters may be included (see Fig. 6.15).

The chart in Fig. 6.14 has been derived for slabs with roughly equal spans in both directions. It will be conservative for non-square grids if the span–depth ratio is based on the larger span.

Example 6.2
Check the punching shear strength of the slab around an internal column

supporting a 225 mm thick flat slab having 6 m spans in both directions. The column is 300 × 400 mm and the design shear force established from analysis of the slab, which has 6·5 m spans in both directions and supports a design ultimate load of 9 kN/m², is 400 kN. Design of the slab for flexure gave an average value for the reinforcement ratio of 0·0077. The characteristic concrete strength is 30 N/mm².

Assuming 20 mm cover and 12 mm dia. bars gives the effective depth as 225 − 20 − 6 = 199 mm. *Table 4.8* gives the basic design shear strength as 0·34 N/mm² for 30 N/mm² concrete. The critical perimeter is

$$2(300+400) + 199 \times 3 \times 3.1416 = 3276 \text{ mm}$$

The effective design shear force is 400 × 1·15 = 460 kN. The shear capacity of the critical perimeter is thus

$$V_{Rd1} = 0.34(1.6-0.2)(1.2+40 \times 0.0077) \times 3276 \times 199/1000 \text{ kN}$$
$$= 469 \text{ kN}$$

This exceeds 460 kN, hence no shear reinforcement is needed.

The same example can be solved using the chart in Fig. 6.14.

(a) $b/d = 300/199 = 1.51$, $h/d = 400/199 = 2.01$

Entering the chart at $b/d = 1.51$, reading across to $h/d = 2.01$ and then down to the horizontal axis gives the perimeter as about $16.5d = 3283$ mm.

(b) Dropping down to the line for $f_{ck} = 30$, then across to the line for $L/d = 6500/199 = 32.66$, and then up to the horizontal axis gives a design load that can be withstood without shear reinforcement as 9·4 kN/m². Since this is greater than 9 kN/m², shear reinforcement is not needed. This corresponds very closely to the result achieved when the calculation was carried out fully.

Had the ultimate load on the slab been 12 instead of 9 kN/m², the effective design shear force would have been 613 kN and shear reinforcement would have been needed. The calculation for this would be as follows.

The shear capacity of the concrete V_{Rd1} is 469 kN. $1.6 V_{Rd1} = 750$ kN. This exceeds the design punching shear force of 613 kN, and hence reinforcement may be provided to support the excess shear above the capacity of the concrete. Thus

$$A_{sw} f_{yd} = 1000(613 - 469) = 144\,000$$

Assuming an f_{yk} value of 460 gives $f_{yd} = 400$. Hence the required area of stirrups is 144 000/400 = 360 mm².

This is a very small amount of reinforcement, and the amount actually supplied may be governed by the minimum reinforcement rules in *clause 5.4.3.3*. This is 0·07%. The interpretation of the rule for punching shear is not clear, but a reasonable approach seems to be to interpret *equation (5.16)* to give

Clause 5.4.3.3

$$A_{sw} = \rho_w u \times 0.75d$$

This gives $A_{sw} = 0.0007 \times 3276 \times 0.75 \times 199 = 342$ mm². This is just less than 360, hence 360 mm² remains appropriate. In order to detail a reasonable arrangement of stirrups, it will almost certainly be necessary to use a considerably larger area of reinforcement.

A check should now be carried out on a perimeter a further $0.75d$ from the face of the column to ensure that no further reinforcement is needed. The length of the next perimeter is $2 \times (300 + 400) + 3.142 \times 4.5 \times 199 = 4213$ mm. The shear capacity of this perimeter is $4213 \times 469/3276 = 603$ kN. This is still slightly below 613 kN, but the load within the perimeter could be deducted. This would reduce the design shear to below 600 kN, so it may be assumed that no further shear reinforcement is needed.

6.5. Torsion

6.5.1. Introduction

In normal building structures, torsion generally arises as a secondary effect and specific calculations are not necessary. Torsional cracking is generally adequately controlled by reinforcement provided to resist shear. Even when torsion occurs, it rarely controls the basic sizing of members and torsion design is often a check calculation after the members have been designed for flexure. Also in some of these cases, the loading that causes the maximum torsional moments may not be the same as that which induces the maximum flexural effects. In some cases, reinforcement provided for flexure and other forces may prove adequate to resist torsion.

It is important to recognize the following two basic types of torsion.

(a) Equilibrium torsion, which is essential for the basic stability of the element or structure, (e.g. a canopy cantilevering off an edge beam (see Fig. 6.15(a)). Equilibrium torsion is clearly a fundamental design effect which cannot be ignored. Structures in which torsion design should be carried out include beams curved on plan, helical staircases and box girders.

(b) Compatibility torsion, which arises in monolithic construction when compatibility of deformation of the connected parts has to be maintained, e.g. edge beams in a normal framed building with slabs and beams framed in on one side only. In these cases the torsional restraint can be released without causing a collapse of the structure; however, at serviceability, cracking may occur in the absence of sufficient reinforcement (see Fig. 6.15(b)).

6.5.2. Evaluation of torsional moments

When torsion has to be taken into account in the design of framed structures, the torsional moments may be obtained on the basis of an elastic analysis using torsional rigidity ($G \times C$) calculated from

$$G = 0.42 E_{cm}$$

$C = 0.5 \times$ St Venant torsional stiffness for a plain concrete section

(0.5 has been justified by test results).

For a rectangular section, the St Venant torsional stiffness $= k_1 b^3 h$, where k_1 is obtained from Table 6.1, b is overall breadth ($<h$) and h is overall depth. Non-rectangular sections should be divided into series of rectangles and their torsional stiffness should be summed up. The division should be carried out so as to maximize the stiffness.

6.5.3. Design verification

In design it should be verified that

$$T_{sd} \leq T_{Rd1}$$

If this is not satisfied, the section should be reinforced such that

SHEAR, PUNCHING SHEAR AND TORSION

Table 6.1. Values of k_1 for a rectangular section

d/b	k_1
1	0·14
1·5	0·20
2	0·23
3	0·26
5	0·29
>	0·33

$$T_{sd} \leq T_{Rd2}$$

T_{sd} is the design torsional moment at ultimate limit state, T_{rd1} is the maximum torsional moment that can be resisted by the notional compression struts in the concrete, and T_{Rd2} is the maximum torsional moment that can be resisted by the reinforcement.

EC2 gives expressions for T_{Rd1} and T_{Rd2} which are discussed below. When a section requires reinforcement for torsion, it should be in the form of closed links and longitudinal reinforcement. All members should be provided with the minimum shear reinforcement recommended in *Table 5.5*. *Clause 4.3.3.1(6)*
Clause 4.3.3.1(7)

Clause 5.4.2.2(5)

In EC2, the torsional resistance of sections is calculated on the basis of a thin-walled closed section. Solid sections are replaced by equivalent thin-walled sections.

EC2 provides expressions for resistance assuming inclined concrete struts. The formulae are reproduced here without explanation, but modified to suit design requirements.

(a) Limiting torsional resistance, above which the section should be reinforced   *Clause 4.3.3.1(6)*

$$T_{Rd1} = 2\nu f_{cd} t A_k / (\cot\theta + \tan\theta)$$

(b) When $T_{sd} > T_{Rd1}$ reinforcement should be provided as follows
 (i) Shear reinforcement   *Clause 4.3.3.1(7)*

$$A_{sw}/s = T_{sd}/2A_k f_{ywd} \cot\theta$$

 (ii) Longitudinal reinforcement   *Clause 4.3.3.1(7)*

$$A_{sl} = u_k (f_{ywd}/f_{yld})(A_{sw}/s) \cot^2\theta$$

The terms are defined below (see also Fig. 6.16).

$t \leq A/u \not> $ the actual wall thickness. In the case of a solid section, t denotes the equivalent thickness of the wall. A thickness less than A/u can be used provided

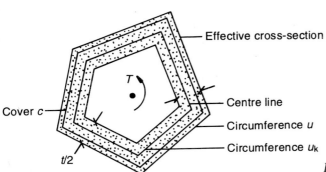

Fig. 6.16. *Notation for torsion*

$T_{sd} \leq T_{Rd1}$, where T_{Rd1} is determined as noted above. A thickness of less than twice the cover c to longitudinal bars is not allowed. u is the outer circumference of the section, A is the total area of the cross-section within the outer circumference, including inner hollow areas, and A_k is the area enclosed within the centre line of the thin walled cross-section, including inner hollow areas.

$$\nu = 0.7(0.7 - (f_{ck}/200)) \not< 0.35 \ (f_{ck} \text{ in N/mm}^2)$$

This value applies if there are stirrups only along the outer periphery of the member. If closed stirrups are provided in both sides of each wall of the equivalent hollow section or in each wall of a box section, ν can be assumed to be $(0.7 - f_{ck}/200) \not< 0.5$.

$$\boxed{0.4} \leq \cot\theta \leq \boxed{2.5}$$

u_k is the circumference of the area A_k, s is the spacing of the stirrups, f_{ywd} is the design yield strength of the stirrups, f_{yld} is the design yield strength of longitudinal reinforcement A_{sl}, A_{sw} is the cross-sectional area of the bars used as stirrups, and A_{sl} is the required additional area of the longitudinal steel for torsion.

T_{Rd1} for a solid rectangular section with shear reinforcement on the outer periphery may be deduced from the general expression in EC2 as follows

$$T_{Rd1} = 2\nu f_{cd} k_2 b^3 / (\cot\theta + \tan\theta)$$

When $\theta = 45°$ this simplifies further to

$$T_{Rd1} = \nu f_{cd} k_2 b^3$$

where b is the breadth of the section ($<d$); k_2 is obtained from Table 6.2 for various (d/b) ratios νf_{cd} is obtained for various values of f_{ck} assuming $\gamma_c = 1.5$.

6.5.4. Combined torsion and bending

As well as the guidance on a general approach, a simplified procedure is given in EC2. Generally this is perfectly adequate in practice. The procedure is as follows.

Clause 4.3.3.2.2(1)
Clause 4.3.3.2.2(2)

(a) Calculate the bending and torsional moments that occur at a section simultaneously.
(b) Calculate the longitudinal reinforcement required for each action effect separately.
(c) In the flexural tension zone, add the torsion reinforcement to that required to resist flexure.
(d) In the compression zone, no additional reinforcement for torsion is necessary if the tensile stress caused by torsion is less than the compressive stress caused by flexure. The tensile stress may be taken as $A_{sl}f_{yld}/A_e$, where A_e is the effective cross-sectional area of the notional hollow section.

Table 6.2. Values of k_2

d/b	k_2	f_{ck}:N/mm²	νf_{cd}:N/mm²
1	0.141	20	5.6
2	0.367	25	6.7
3	0.624	30	7.7
4	0.864	35	8.6
		40	9.3

(e) When torsion is combined with a large bending moment, calculate the principal stress caused by the mean longitudinal compression in flexure and a tangential stress equal to $T_{sd}/(2A_k t)$. The principal stress should be less than αf_{cd} where α is the coefficient defining the compression stress block (*clause 4.2.1.3.3(11)*).

Clause 4.2.1.3.3(11)

6.5.5. Combined torsion and shear

A circular interaction formula is used in EC2. Some other codes, e.g. BS8110, use a linear interaction, which is more conservative.

Clause 4.3.3.2.2(3)
Clause 4.3.3.2.2(4)
Clause 4.3.3.2.2(5)

$$(T_{sd}/T_{Rd1})^2 + (V_{sd}/V_{Rd2})^2 \leq 1$$

(see Fig. 6.17).

The reinforcement required for shear and torsion may be calculated separately and added. It is important to note that the value of the inclination of concrete struts θ must be the same for shear and torsion design.

In general, designed reinforcement should be provided when $T_{sd} > T_{Rd1}$. However, EC2 exempts approximately rectangular sections under the conditions

$$T_{sd} \leq V_{sd} b_w / 4 \cdot 5$$
$$V_{sd}[1 + (4 \cdot 5 T_{sd} / V_{sd} b_w)] \leq V_{Rd1}$$

where V_{sd} is the design shear force and b_w is the minimum width of section over the effective depth.

These conditions are shown in Fig. 6.18.

6.5.6. Detailing

(a) Shear reinforcement for torsion should be in the form of closed links, which should be effectively anchored.

Clause 5.4.2.3

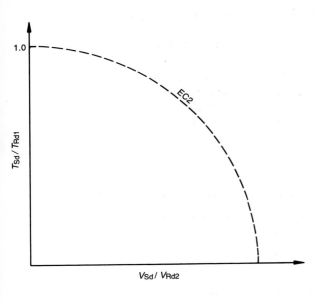

Fig. 6.17. Maximum combined shear and torsion

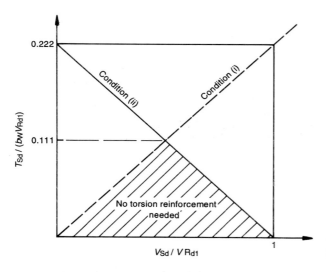

Fig. 6.18. Conditions where no torsion reinforcement is needed

(b) In addition to the requirement for links in beams (see Chapter 10), the longitudinal spacing of torsion links should not exceed $(u_k/8)$. For the definition of u_k, see section 6.5.3 above.

(c) The longitudinal bars should be so arranged that there is at least one bar in each corner. Any other bars required may be distributed around the periphery inside the links at spacings not exceeding 350 mm.

CHAPTER 7

Slender columns and beams

7.1. Scope
This chapter covers the material in *clause 4.3.5*, 'Ultimate limit states induced by structural deformation (buckling)'. Also covered is the material in *Appendix 3*, which provides supplementary information on this subject, including a set of flow charts which will be found very useful as a guide to the clauses.

Clause 4.3.5
Appendix 3

7.2. Background to design of columns for slenderness effects
When an element is subjected to an axial load combined with a moment, it will deflect. This deflection will increase the moment at any section in the element by an amount equal to the axial force multiplied by the deflection at that point. This extra moment will cause the resistance of the element to be reduced below that calculated when deflections are ignored. In many, if not most, practical situations the effect of deflections is so small that it can be ignored. Indeed, in many practical cases the effect of deflections on the design of a member is not merely insignificant but actually nil. How this comes about is explained below. A slender column is one where the influence of deflections on the ultimate strength is significant and must be taken into account in the design.

Slenderness effects in struts have been the subject of theoretical and experimental study from the earliest days of the science of structural mechanics; the Euler equation for the buckling strength of a strut must be well known to every engineer. The Euler equation deals with the failure by instability of a strut made of infinitely strong, elastic material. Failure occurs due not to failure of the material, but because the deflection under the critical buckling load becomes infinite. The problem that the designer faces in the design of reinforced concrete members is very rarely one of classical buckling: failure of the material occurs at a lower load than would be predicted if the deflections were ignored, due to the moments being higher than predicted. The other critical difference from classical buckling theory is that reinforced concrete is definitely not elastic at the ultimate limit state.

In order to understand what the design rules given in EC2 are aiming to achieve, it is necessary to have an idea of the forms of deflection that may occur. Two modes of deflection are normally possible: sidesway of the whole structure and deflection of a single column without sidesway. These are shown in Fig. 7.1: the two modes give very different forms of bending moment diagram in the columns. In the sidesway mode, note that all the columns within the storey height are subjected to the same deflection.

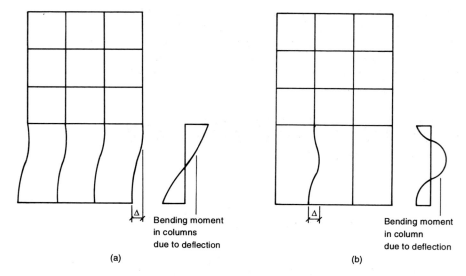

Fig. 7.1. Modes of deflection of columns in a structure: (a) sidesway of whole structure; (b) deflection of a single column

Code rules dealing with slenderness effects commonly include three basic steps

(a) establish which modes of deflection are likely
(b) establish whether or not slenderness effects are likely to be significant (i.e. whether or not the structure or member considered is slender)
(c) if the slenderness effects are likely to be significant, design to make allowance for them.

This is the approach adopted by EC2, and the basic principles are introduced here. Detailed guidance on the use of the provisions is given later in this chapter.

In order to establish the deflection modes that should be considered, EC2 classifies structures as either sway or non-sway and as either braced or unbraced. A sway structure is one where sidesway is likely to be significant (as in Fig. 7.1(a)); in non-sway structures, sidesway is unlikely to be significant. 'Significance' is defined in this context as a situation where lateral displacement of the ends of the columns increases the critical bending moments by more than 10% above the moment calculated ignoring the displacements. Basically, whether a structure is sway or non-sway will depend on its stiffness against lateral deformation. In most cases a structure can be classified by inspection, but where the classification is not obvious, *Appendix 3.2* gives formulae by which it can be assessed more rigorously. These are repeated here for convenience.

Appendix 3.2

A framed structure with bracing elements may be classified as non-sway if the lateral stiffness of the bracing elements satisfy the following criteria

$$\text{for } n \leq 3, \; k \leq 0.2 + 0.1n$$

$$\text{for } n \geq 4, \; k \leq 0.6$$

where

$$k = h_{\text{tot}} \sqrt{(F_v / E_{cm} I_c)}$$

n is the number of storeys, h_{tot} is the total height of the structure in metres, $E_{cm} I_c$ is the sum of the nominal flexural stiffnesses of all vertical bracing elements acting in the direction considered, and F_v is the sum of all vertical loads carried by the bracing elements under service conditions.

The principle behind these equations can easily be established. Consider the bracing elements of a structure as a cantilever of height h_{tot} loaded at the top by a horizontal load H and a vertical load N. The deflection of the cantilever can be considered to be made up of two components: a deflection due to the horizontal load and a further deflection due to the action of the vertical load. The deflection due to the horizontal load is given by

$$a_h = Hl^3/3EI$$

The additional deflection due to the vertical load can be assessed if it is assumed that the deflected shape of the column is sinusoidal as

$$a_v = Nl^2 a_{tot}/10EI$$

a_{tot} is the total deflection, given by

$$a_{tot} = a_v + a_h = Hl^3/3EI + Nl^2 a_{tot}/10EI$$

Hence

$$a_{tot} = 10Hl^3/[3EI(1 - Nl^2/10EI)]$$

In the limit, for a structure to be classed as non-sway

$$Na_{tot} \leq 0.1Hl$$

Hl is the first-order moment, and the deflection should not affect this by more than 10%. Hence

$$10NHl^2/[3EI(1 - Nl^2/10EI)] \leq Hl/10$$

This can be simplified to

$$l\sqrt{(N/EI)} \leq 0.54$$

which has the same form as the limits given in EC2 and quoted above. The difference in the values on the right-hand side of the inequality arises from three sources:

(a) the horizontal load is likely to be more uniformly distributed up the height of the building rather than concentrated at the top, as assumed in the derivation above
(b) the vertical load will not be applied at the top of the column, but incrementally over the height at each floor level
(c) the interconnection between bracing elements at each floor level will modify the deflected shape, which will increasingly deviate from the deflected shape of a cantilever as more floors and more bracing elements are introduced.

When properly taken into account, these factors will lead to the limits given in EC2.

A frame structure without bracing elements may be considered non-sway if each vertical element in the frame that resists more than 70% of the average axial force in all the vertical elements has a slenderness ratio less than the greater of 25 or $15/\sqrt{v_u}$.

A braced structure is one that contains bracing elements. These are vertical elements (usually walls) which are so stiff relative to the other vertical elements that they may be assumed to attract all horizontal forces. A bracing element or system of bracing elements should be so stiff that it will attract, and transmit to the foundations, at least 90% of all horizontal loads applied to the structure (*clause 4.3.5.3.2*). As with sway/non-sway, classification can normally be by inspection, but clearly it could also be by calculation without undue difficulty.

In general, where the bracing elements in a braced structure consist of shear

Clause 4.3.5.3.2

walls or cores, braced structures may be assumed to be non-sway without further checking.

The structure having been classified, the following applies.

(a) Non-sway structures: only the deformation of individual columns of the type illustrated in Fig. 7.1(b) need be considered as, by definition, sidesway will be insignificant.

(b) Sway structures
  (i) Braced structures: as mentioned above, braced structures may normally be assumed to be non-sway unless the bracing elements are relatively flexible. If they are, the bracing elements or structure should be analysed for the effects of sidesway but the braced elements within the structure may be assumed to deform as illustrated in Fig. 7.1(b), and sidesway may be ignored.
  (ii) Unbraced structures: these structures should first be designed to take account of sidesway of the whole structure, as illustrated in Fig. 7.1(a), and then each column in turn should be analysed for the possibility of its deforming as illustrated in Fig. 7.1(b).

The second element in the EC2 provisions is to establish whether slenderness effects are likely to be significant or whether they can be ignored. EC2 defines significance as being when slenderness effects increase the first-order moments by more than 10%. It is useful to have this stated as a general principle, but from a practical point of view it is unhelpful, since one can establish that the effect is less than 10% only by doing a full calculation, whereas the objective is to establish when the calculation is unnecessary. Some simplified rules that may be deemed to satisfy this requirement are therefore needed. Like most other codes, EC2 formulates these rules in terms of two parameters: the effective length of the column and the slenderness ratio.

Effective length is a concept used almost universally in the treatment of second-order effects. The effective length can best be defined as the length of a pinned ended strut of the same cross-section as the column considered and which would have the same buckling strength. The concept is illustrated in Fig. 7.2. The effective length is seen to depend on the type of deflection considered and the end fixity of the column. A non-sway column will have an effective length between 0·5

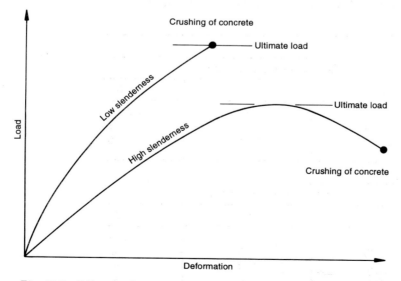

Fig. 7.2. *Effective length*

and 1·0 times the actual clear height of the column; a sway column will have an effective length greater than the clear height.

Only in very rare cases will the effective height exceed twice the overall height. Formulae can be derived to give the effective height as a function of the stiffnesses of the column and the members framing into each end of the column. *Fig. 4.25* provides a nomogram that will enable the calculation of effective lengths. In some cases, for example in computer programs, it may be more convenient to have equations. The equations given below appear in both the UK Code, BS8110, and the US ACI code.

(a) For columns in non-sway frames, the effective length may be taken as the lesser of

$$l_0[0\cdot7 + 0\cdot05(k_a + k_b)] < l_0$$
$$l_0(0\cdot85 + 0\cdot05k_{min}) < l_0$$

(b) For columns subjected to sway, the effective length may be taken as the lesser of

$$l_0[1\cdot0 + 0\cdot15(k_a + k_b)]$$
$$l_0(2\cdot0 + 0\cdot3k_{min})$$

In these equations, l_0 is the clear height of the column between the end restraints, k_a is the ratio of the sum of the column stiffnesses to the sum of the beam stiffnesses at one end of the column, k_b is the ratio of the sum of the column stiffnesses to the sum of the beam stiffnesses at the other end of the column, and k_{min} is the lesser of k_a and k_b.

BS8110 provides a table of approximate values (reproduced here as Table 7.1), derived from the equations given above, which are probably adequate for columns in most normal buildings. In Table 7.1, the definitions of the four end conditions are as follows.

(a) Condition 1: the end of the column is connected monolithically to beams on either side which are at least as deep as the overall dimension of the

Table 7.1. Approximate values for effective length factors (effective length = factor × clear height)

Braced or non-sway columns			
End condition at top	End condition at bottom		
	1	2	3
1	0·75	0·80	0·90
2	0·80	0·85	0·95
3	0·90	0·95	1·00

Sway columns			
End condition at top	End condition at bottom		
	1	2	3
1	1·2	1·3	1·6
2	1·3	1·5	1·8
3	1·6	1·8	—
4	2·2	—	—

column in the plane considered. Where the column is connected to a foundation, this should be of a form specifically designed to carry moment.

(b) Condition 2: the end of the column is connected monolithically to beams or slabs on either side which are shallower than the overall dimension of the column in the plane considered.

(c) Condition 3: the end of the column is connected to members which, while not specifically designed to provide restraint to rotation of the end of the column, will nevertheless provide some nominal restraint.

(d) Condition 4: the end of the column is unrestrained against both lateral movement and rotation (e.g. the free end of a cantilever column in an unbraced frame).

The other factor used to define the slenderness of a column is the slenderness ratio. This is defined as the ratio of the effective length to the radius of gyration. The radius of gyration of a section is defined by

$$i = \sqrt{(I/A)}$$

where i is the radius of gyration, I is the second moment of area of the cross-section, and A is the area of the cross-section.

For a rectangular section, $I = bh^3/12$ and $A = bh$, hence the radius of gyration $= 0.2887h$ where h is the dimension of the section perpendicular to the axis of bending considered. The radius of gyration of a circular section is equal to its actual radius.

Many codes define the slenderness ratio as the ratio of the effective length to the section depth. It is not easy to argue that one method is intrinsically better than the other for reinforced concrete since, as discussed above, the problem is not one of classical buckling and reinforced concrete is not an elastic material. Use of the radius of gyration does, however, have the advantage of giving a reasonable way of dealing with non-rectangular sections.

The limits given in EC2 for the slenderness ratios above which it will be necessary to take account of slenderness effects for non-sway columns are as follows.

(a) If the slenderness ratio is less than the lesser of $\lambda_{min} = 25$ and $15/\sqrt{v_u}$, slenderness effects may be ignored.

(b) If the slenderness ratio is in the range

$$\lambda_{min} < \lambda < 25(2 - e_{01}/e_{02}) \ (= \lambda_{crit})$$

it should be checked that the column ends can withstand a moment of at least $N_{sd}h/20$.

No simple limits are given in EC2 for sway frames, but it seems reasonable to assume that any structure that is classifiable as a sway frame by the rules set out in *Appendix 3.2* will need to be designed for the effects of deformations.

Appendix 3.2

7.3. Design for slenderness effects

Once it has been concluded that a column or structure needs to be designed for slenderness effects, the remaining, and biggest, problem is how this should be done. The rest of this chapter is concerned with this problem.

EC2 provides two basic approaches: rigorous non-linear analysis of the structure and simplified methods for most normal situations. Assumptions are given for the former, but no precise method is specified. These assumptions and the basic approach are introduced below, but it is beyond the scope of this manual to provide a detailed development of a suitable calculation method. This is a matter for experts in computer programming. A simplified method is given in EC2 in detail for the design of single braced columns; this is discussed in detail below. EC2 does not give details of a method for dealing with sway frames, although it does say that the method developed for braced columns may be adapted for use in sway frames, subject to limitations. A suitable method is developed in section 7.3.6 below.

All methods require that the possibility of the structure being accidentally constructed out-of-plumb should be considered. This is allowed for by introducing an accidental eccentricity of the axial load in the case of an isolated column or by considering an accidental inclination from the vertical for a sway frame. *Clause 2.5.1.3* defines the accidental inclination in radians as

Clause 2.5.1.3

$$v = 1/(100\sqrt{l}) \le 1/200$$

where v is the inclination in radians and l is the total height of the structure in metres.

The limit of $1/200$ corresponds to a height of structure of 4 m. For all structures above this height, v will be less than this limiting value. Additionally, where n vertically continuous members act together, v may be reduced by a factor equal to $\sqrt{[(1+(1/n)]/2}$. Braced columns are designed for an accidental additional eccentricity of the longitudinal force given by

$$e_a = v\, l_o/2$$

where l_0 is the effective length of the column.

7.3.1. Assumptions for non-linear analysis and outline of procedure

EC2 gives the following assumptions about the behaviour of the materials and reinforced concrete, which should be used in carrying out an analysis to establish the effects of slenderness.

For *concrete*, the stress–strain diagram shown in Fig. 7.3 (reproduced from *Fig. 4.1*) should be used with design ultimate values taken for f_c and E_c. In assessment of the design values, γ_c may be taken at the reduced value of $1\cdot 35$. Thus

$$f_c = f_{ck}/1\cdot 35$$
$$E_c = E_{cm}/1\cdot 35$$

E_{cm} is given in *clause 3.1.2.5.1 (Equation 3.5)* as

Clause 3.1.2.5.1

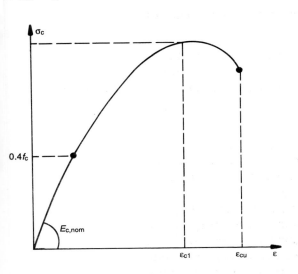

Fig. 7.3. Stress–strain diagram for structural analysis

$$E_{cm} = 9 \cdot 5(f_{ck}+8)^{1/3}$$

In calculating the deformations due to creep, a partial safety factor of $1 \cdot 1$ for indeterminate structures and $1 \cdot 2$ for determinate structures may be applied to the quasi-permanent load. Creep may most conveniently be dealt with by using an effective elastic modulus where

$$E_{cm.eff} = E_{cm}/(1+\phi)$$

where ϕ is the creep coefficient. This may be derived from *clause 2.5.5*. *Clause 2.5.5*

For *reinforcement*, the normal design ultimate stress–strain curve used for section design may be used. It may be useful to introduce the sloping upper branch of the curve, as this may help with convergence problems in the mathematical procedures.

For *reinforced concrete*, plane sections are assumed to remain plane.

The contribution of the concrete in tension in cracked sections may be taken into account using the formula for an effective average steel strain given in *Appendix 2*. This is repeated here for convenience *Appendix 2*

$$\epsilon_{sm} = \epsilon_{smr} + \epsilon_s/\epsilon_s[1-\beta_1\beta_2(\sigma_{sr}/\sigma_s)^2]$$

This equation differs slightly from the approach given in *Appendix 4*. There seems *Appendix 4* no good reason for this, and the equations in *Appendix 4* are more rigorously defensible (see section 8.4 of this manual). The actual practical difference will, however, be small.

For calculation of the behaviour of concrete in compression, the stress–strain curve shown in Fig. 7.3 should be adopted with design values introduced for the concrete strength and the elastic modulus. These are obtained from

$$f_{cd} = f_{ck}/\gamma_c$$
$$E_{cd} = E_{cm}/\gamma_c$$

The effects of creep should be taken into account where they increase the first-order bending moments by more than 10%. No method is given that will permit assessment of whether or not the effect will exceed 10% prior to calculation: however, a rough assessment could be made on the basis of the following argument for isolated columns. *Clause 4.3.5.5.2(2)* implies that slenderness effects will not *Clause* be significant where the slenderness ratio is less than λ_{crit}. From this it can be *4.3.5.5.2(2)* inferred that slenderness effects, *including creep*, do not increase the first-order moments by more than 10% where $\lambda < \lambda_{crit}$. If it is assumed that creep does not account for more than 50% of the deflection and that the deflection is proportional to the square of the slenderness ratio, then creep will not cause an increase in the first-order moments greater than 10% provided that

$$\lambda < 35(2-e_{01}/e_{02})$$

If it is necessary to take creep into account, then various possible methods of doing this are suggested, i.e. to

(a) use the information on creep set out in *clause 2.5.5* *Clause 2.5.5*
(b) modify the stress–strain curve for the concrete
(c) increase the additional eccentricity e_a.

No clear procedure is set out in EC2 for carrying out any of these options. It seems, however, that a practical approach may be to use elements of all three as described below.

The creep deformation is assessed under only the design quasi-permanent load, not the design ultimate load. This is obtained by applying a γ_F of $1 \cdot 1$ for indeterminate structures or 1·2 for determinate structures to the quasi-permanent load. It is suggested that the creep information given in *Table 3.3* in *clause 3.1.2.5.5* be used to establish a creep coefficient ϕ. The strains in the concrete stress–strain diagram are then multiplied by $(1 + \phi)$ to give a modified stress–strain diagram applicable for loadings up to the design quasi-permanent load. The deflected state of the structure is assessed under the action of the quasi-permanent loads on the basis of both the short-term stress–strain diagram and the creep-modified diagram. The difference between these two calculations is the creep deflection. This creep deflection is added to the accidental deflection e_a, and the strength of the structure is assessed on the basis of the short-term stress–strain curve.

Clause 3.1.2.5.5

Where a rigorous approach is used for the assessment of second-order effects, lower partial safety factors may be adopted. *Appendix 3.1* suggests that the partial safety factors on the loads may be reduced by 10% (say from 1·5 to 1·35 and from 1·35 to 1·2) for structures with an overall height in excess of 22 m. The partial factor on the concrete γ_c may be reduced to 1·35.

Appendix 3.1

It is not possible in a book of this type to give a detailed description of methods of calculation for slender structures. This can be done only by computer, and the development of suitable programs is not straightforward. The basic steps used in typical programs can, however, be outlined as follows.

(a) Select a value of axial load.
(b) Guess a deflected shape for the structure. A sine curve is usually a good starting point.
(c) From the axial load, the first-order eccentricities and the assumed deflected shape, calculate the bending moment diagram for the structure.
(d) Using this bending moment diagram, calculate the deflected shape of the structure.
(e) If the deflected shape calculated in (d) differs significantly from that assumed in (b), choose a modified deflected shape and return to step (c).
(f) Increase the axial load and repeat the process from step (b). In this case, a reasonable estimate of the deflected shape can be assessed by extrapolation from the deflections at the previous load stage provided that the load increment is not too big.
(g) The failure load corresponds to the condition where it is impossible to establish a deflected shape that will permit the loads to be supported.

The assessment of slenderness effects by the use of non-linear analysis is a process of some complexity, and should not be attempted except in special cases. For almost all normal practical situations, simplified methods must be used.

7.3.2. Simplified method

A simplified method is given for the analysis of isolated columns in *clause 4.3.5.6.3*, where it is called the 'model column method'. The principles behind this method merit a brief description.

Clause 4.3.5.6.3

If a rigorous non-linear analysis is carried out for a slender column, it will result in a load–deflection curve of a similar form to one of the curves in Fig. 7.2, depending on the slenderness ratio. The object of a rigorous analysis is to calculate such curves and hence find the maximum load capacity. The ideal simplified method would aim to establish this ultimate load, or something close to it, in a single calculation without the need to calculate the full load–deflection response. What the EC2 method aims to do is to predict the deflection at which failure of the concrete will begin (i.e. maximum compressive strain $= \epsilon_u$). It is seen from Fig. 7.2 that, if this point can be established, it will correspond either to the actual ultimate load or to a conservative estimate of the ultimate load. Such a method will therefore, in principle, give a lower-bound estimate of the strength. The calculation of the ultimate deflection is carried out as follows for a pinned ended strut.

The ultimate curvature of the section is estimated: for a balanced section, the strain at the compression face of the section is $0 \cdot 0035$, while the strain in the reinforcement is given by f_{yd}/E_s. From this, the curvature can be written as

$$1/r_u = [0 \cdot 0035 + (f_{yd}/E_s)]/d$$

For axial loads above the balance point, the curvature will be less than this, as can be seen from Fig. 7.4. To allow for this, the balanced curvature is multiplied by a factor K_2, given by

$$K_2 = (N_{ud} - N_{sd})/(N_{ud} - N_{bal})$$

where N_{ud} is the design ultimate axial load capacity of the section ($= f_{cd}A_c + f_{yd}A_s$), N_{sd} is the design axial load on the column, and N_{bal} is the design load capacity of a balanced section (for a symmetrically reinforced rectangular section, this may be taken as $0 \cdot 4 f_{cd} A_c$).

It will be seen that, for a balanced section, $K_2 = 1$, and that K_2 reduces to zero as the load approaches the axial load capacity of the section. This approach provides an approximate estimate of the ultimate curvature. A better method might be to define K_2 as ϵ_u/x, although this may not be as convenient.

Figures 7.5 and 7.6 can be used to obtain values of K_2 for symmetrically reinforced rectangular sections.

Calculate the ultimate deflection: this may be done by assuming that the column deflects in the form of a sine curve. Thus, if it is assumed that the curvature is proportional to the moment, which is equal to the deflection multiplied by the axial load

$$a = \iint 1/r_u \sin(\pi x/l_0)$$
$$= l_0^2/\pi^2 (1/r_u) \sin(\pi x/l_0)$$

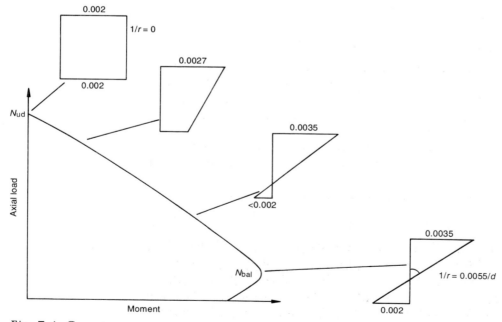

Fig. 7.4. *Curvature*

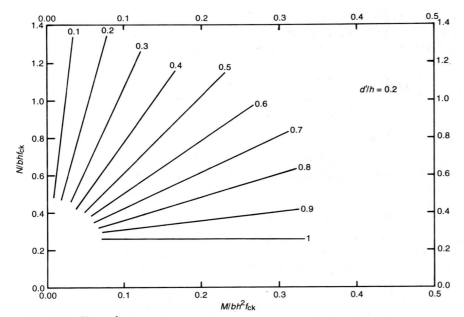

Fig. 7.5. K_2 values

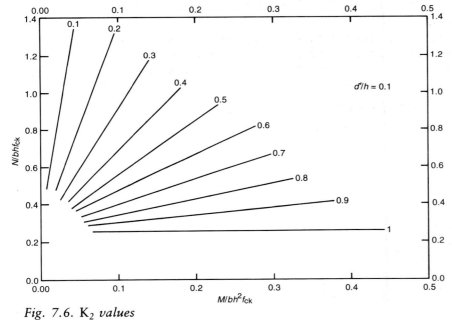

Fig. 7.6. K_2 values

at mid-height, $x/l_0 = 0.5$ and hence

$$a \approx 0.1 \, (1/r_u) l_0^2$$

The ultimate deflection at mid-height of the column can thus be calculated. On the one hand, the estimate is likely to be conservative since the assumption of a sinusoidal variation of curvature over the height of the column will be conservative due to the non-linear behaviour of reinforced concrete. On the other hand, some of this conservatism will be counterbalanced by the coefficient K_2, giving an unconservative estimate of the curvature in some cases.

In EC2, the formulae for the curvature and the deflection are not quite as set out above. A slightly simplified equation is derived for the curvature by assuming that the neutral axis depth for a balanced section is $0.55d$. The curvature can then be written as

$$1/r_u = K_2 \epsilon_{yd}/0.45d$$

A further coefficient K_1 has been introduced into the deflection formula to give a reduced influence to slenderness effects at low slenderness ratios. Thus

$$e_2 = a = 0.1 K_1 l_0^2 (1/r_u)$$

where

$$K_1 = \lambda/20 - 0.75 \text{ for } 15 \leq \lambda \leq 35$$
$$= 1 \text{ for } \lambda > 35$$

For most normal columns, slenderness effects will not have to be considered until the slenderness ratio is well above 35 since λ_{crit} is likely to exceed 35. K_1 can thus normally be taken as 1.0.

Table 7.2 gives values of $e_2/K_1 K_2 d$ as a function of $(l_0/d)^2$ for various grades of reinforcement.

Define the design moments in the structure: the analysis so far considers the deformation of a pinned ended strut, but this is not the normal configuration of a column in a building. A normal column built monolithically into a structure at top and bottom will deform, and be subjected to moments like those shown in Fig. 7.7. It will be seen that the section of the column between the points of contraflexure in the final state of the column may be considered to be a pinned ended strut equivalent to that for which the analysis was carried out. The distance between the points of contraflexure is the effective length of the column. The estimation of this length is covered above. The maximum moment due to deflection will be seen to occur at mid-height of the effective column. This will normally be somewhere close to mid-height of the real column. Clearly, the total moment to which the critical section is subjected is made up of the maximum moment due to the deflection plus the first-order moment at this height, plus any allowance

Table 7.2. Values of $e_2/K_1 K_2 d$

l/d	Values of eccentricity/($K_1 K_2 d$) for f_y =				
	220	400	450	460	500
10	0.0213	0.0386	0.0435	0.0444	0.0483
12	0.0306	0.0557	0.0626	0.0640	0.0696
14	0.0417	0.0757	0.0852	0.0871	0.0947
16	0.0544	0.0989	0.1113	0.1138	0.1237
18	0.0689	0.1252	0.1409	0.1440	0.1565
20	0.0850	0.1546	0.1739	0.1778	0.1932
22	0.1029	0.1871	0.2104	0.2151	0.2338
24	0.1224	0.2226	0.2504	0.2560	0.2783
26	0.1437	0.2613	0.2939	0.3004	0.3266
28	0.1666	0.3030	0.3409	0.3484	0.3787
30	0.1913	0.3478	0.3913	0.4000	0.4348
32	0.2177	0.3957	0.4452	0.4551	0.4947
34	0.2457	0.4468	0.5026	0.5138	0.5585
36	0.2755	0.5009	0.5635	0.5760	0.6261

SLENDER COLUMNS AND BEAMS

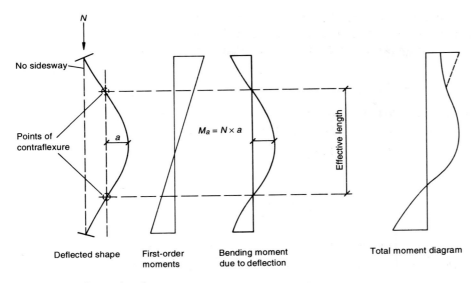

Fig. 7.7. Braced column

for accidental effects. A reasonable estimate of the first-order moment near mid-height of the column is given by

$$M_o = 0\cdot 6 M_{max} + 0\cdot 4 M_{min}$$

where M_{max} and M_{min} are the numerically greater and lesser end moments respectively. This relation, written in terms of eccentricities, is included in EC2 (*clause 4.3.5.6.2*) together with a further precautionary limitation that the moment should not be less than $0\cdot 6 M_{max}$.

Clause 4.3.5.6.2

It will be seen from Fig. 7.7 that as well as the moment near mid-height being modified by the deflections, the end moments are affected. The numerically larger end moment will actually be reduced, while the numerically smaller end moment will be increased. Thus, rigorously, there are three possible conditions that should be considered in establishing the critical conditions for section design.

(a) The numerically larger end moment, assuming that no deflection occurs. In most columns of intermediate slenderness, this is the critical design condition and, although the deflection influences the moments at mid-height of the column, it does not influence the design. In EC2 the range of columns where this can be relied upon to be the case are those that have slenderness ratios laying between λ and λ_{crit}. Nevertheless, columns may frequently occur where the design eccentricity given by *equation (4.64)* is less than the larger end eccentricity. Clearly, the larger end eccentricity must be taken as the design value in this case.

(b) The moment at mid-height, allowing for the ultimate deflection. This is the condition covered explicitly in EC2.

(c) The numerically smaller end moment increased by the effects of deflections. The increase will be less than that corresponding to the second-order eccentricity at mid-height. The UK Code, BS8110, assumes that the second-order moment at this end of the column is equal to half the second-order moment at mid-height. EC2 does not require a check at this point and, in fact, it is rarely critical.

Example 7.1
Calculate the design ultimate moment for which the column shown in Fig. 7.8(a) should be designed. The column is in a braced frame. The stiffness of the beams and columns framing into the column considered can all be assumed to have the same stiffness L/EI as the column.

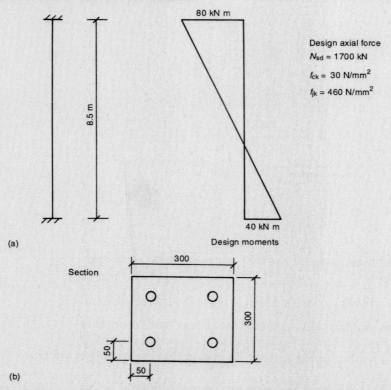

Fig. 7.8. *Example of application of the simplified column method*

(a) Calculate the effective height: in *Fig. 4.25*, $K_a = K_b = 1 \cdot 0$ and hence $\beta = 0 \cdot 77$. The effective height is thus $0 \cdot 77 \times 8 \cdot 5 = 6 \cdot 55$ m

(b) Calculate the slenderness ratio

$$\text{slenderness ratio} = \text{effective height/radius of gyration}$$
$$= 6550/0 \cdot 2887 \times 300 = 75 \cdot 63$$

(c) Calculate slenderness limits

$$\lambda_{min} = \text{greater of 25 and } 15/\sqrt{\nu}$$
$$= 15/[1700 \times 1000/300_2/(30/1.5)]^{0 \cdot 5} = 15.46$$

hence $\lambda_{min} = 25$

$$\lambda_{crit} = 25[2-(-50)/80] = 65 \cdot 6 \ [equation \ (4.61)]$$

The column is thus slender and, since the slenderness exceeds λ_{crit}, specific measures to deal with slenderness are necessary.

(d) Calculate e_o

M_o = greater of $0 \cdot 6 \times 80 - 0 \cdot 4 \times 50 = 28$ and $0 \cdot 4 \times 80 = 32$

Hence

$$M_o = 32 \text{ and } e_o = 32 \times 10^6/1\,700\,000 = 18 \cdot 8 \text{ mm}$$

(e) Calculate accidental eccentricity e_a (assume $\nu = 1/200$)
$$e_a = 1/200 \times 6 \cdot 55 \times 1000/2 = 16 \cdot 4 \text{ mm}$$

(f) Calculate the second-order eccentricity e_2
$$l_0/d = 6550/250 = 26 \cdot 2$$

From Table 7.2, $e_2/dK_1K_2 = 0 \cdot 3009$. Since $\lambda > 35$, $K_1 = 1$, hence
$$e_2 = 0 \cdot 3009 \times 250 \times K_2 = 75 \cdot 2 K_2$$

A value for K_2 can be found from Fig. 7.6 by iteration as follows
$$N/bhf_{ck} = 1700 \times 1000/300/300/30 = 0 \cdot 63$$

A first estimate of the moment may be obtained by assuming $K_2 = 1$. This gives
$$M/bh^2f_{ck} = 1700 \times 1000 \times (75 \cdot 2 + 16 \cdot 4 + 18 \cdot 8)/300^3/30 = 0 \cdot 23$$

From Fig. 7.6, this gives K_2 as $0 \cdot 66$. This leads to a modified total moment of
$$M/bh^2f_{ck} = 1700 \times 1000 \times (75 \cdot 2 \times 0 \cdot 66 + 16 \cdot 4 + 18 \cdot 8)/300^3/30 = 0 \cdot 178$$

From Fig. 7.6, this leads to a second estimate of K_2 of $0 \cdot 60$ and a modified value of M/bh^2f_{ck} of $0 \cdot 169$. A final round of iteration gives $K_2 = 0 \cdot 59$ and $M/bh^2f_{ck} = 0 \cdot 166$.

Figure 5.9 may now be used to obtain a value for pf_{yk}/f_{ck} of $0 \cdot 65$, and hence a steel area of 3815 mm^2.

7.3.3. Other factors

The sections above are concerned exclusively with cases where the second-order effects act about the same axis as the first-order moments and where only uniaxial moments act. While this will frequently be the case, it will by no means always be so. Two conditions may need to be considered: first, cases where the column may deflect about the minor axis even though the first-order moments act only about the major axis, and second, cases of biaxial bending. These two conditions are now considered in turn.

Clause 4.3.5.5.3(p3) states that, for columns bent dominantly about one principal axis, the possibility of failure about the other principal axis should be considered. The following method is given, which generally avoids the necessity to design for biaxial effects.

Clause 4.3.5.5.3(p3)

The column should be designed to resist two possible cases of loading:

(a) bending about only the axis subjected to the first-order moments — in this case the design eccentricity will be given by
$$e = e_0 + e_a + e_2$$

(b) bending about the other axis — in this case the design eccentricity will be given by
$$e = e_a + e_2$$

Provided the first-order eccentricity e_0 is less than $0 \cdot 2h$, checks (a) and (b) above may be carried out independently. Where e_0 exceeds $0 \cdot 2h$, the second check should be carried out using a reduced section size. The dimension parallel to the second axis of bending is reduced to the part of the section that will be in compression under the action of the axial load and the first-order plus accidental eccentricities acting about the first axis. This is illustrated in Fig. 7.9. Fig. 7.10

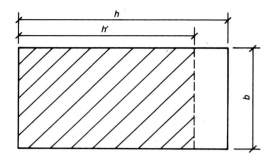

Fig. 7.9. Reduced h' for separate check for minor axis deformation

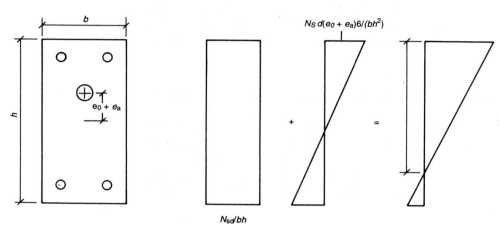

Fig. 7.10. Calculation of reduced h

illustrates the procedure for calculating the reduced dimension h'. It will be seen that the dimension y from the centroid of the section to the point of zero stress is given by

$$N_c/A_c - N_c(e_0 + e_a)y/I = 0$$

For a rectangular section, this can be reduced to the following formula for h'

$$h' = h[0\cdot 5 + (h/12)(e_0 + e_a)]$$

If a means for designing for biaxial bending is readily available, then an alternative to the use of a reduced dimension h' in the second check would be to design for the following case of biaxial bending: $e_0 + e_a$ about the first axis combined with $e_a + e_2$ about the other axis.

For the case of columns where there are significant moments about both axes simultaneously, the rigorous requirement would be to design for the following two load cases:

(a) $e_{0z} + e_{az} + e_{2z}$ about the major axis combined with $e_{0y} + e_{ay}$ about the minor axis

(b) $e_{0z} + e_{az}$ about the major axis combined with $e_{0y} + e_{ay} + e_{2y}$ about the minor axis.

EC2 does, however, define situations where a simplified approach may be used, whereby checks on the basis of uniaxial bending are carried out for the two axes independently. These are cases where the first-order bending is close to being uniaxial at the critical section for buckling. This condition is defined in *clause 4.3.5.6.4* as being when either $e_{0z}b/e_{0y}h \leq 0\cdot 2$ or $e_{0z}b/e_{0y}h \geq 5$.

Clause 4.3.5.6.4

In addition, if $e_{0z}/h > 0.2$, then, when one is checking for bending about the minor axis, a reduced value for the larger dimension of the section h' should be used, given by

$$h' = h[0.5 + h/12(e_{0z} + e_{az})]$$

7.3.4. Simplified method for the design of sway frames

EC2 does not directly give any simplified procedure for the design of sway frames but it does say in *Appendix 3.5(2)* that 'the simplified methods defined in *clause 4.3.5* may be used instead of a refined analysis, provided that the safety level required is ensured'.

Appendix 3.5(2)
Clause 4.3.5

The method proposed here is adapted from the model column method described in 7.3.3, with adjustments to allow for the particular conditions obtaining in sway frames. A fuller discussion of the method and its justification are given in the Appendix.

The basic assumptions of the method are that

(a) the overall deflection of all columns at any given level within a frame is the same
(b) failure of the structure is deemed to have occurred when the first structurally significant column in any storey (the critical column) reaches its ultimate deflection
(c) the ultimate deflection is given by *equations (4.69)–(4.73)* (i.e. the equations are the same as are used for braced columns, but the effective lengths will be larger)
(d) the moment–deflection curve for a column is assumed to be a parabola with its maximum at the ultimate moment and ultimate deflection.

These assumptions are used to develop the following design procedure.

1. Calculate the critical ultimate deflection. It will often be possible to identify the critical column by inspection; otherwise the ultimate deflection of each column should be calculated assuming that they are independent isolated columns subjected to sidesway. The critical deflection is the smallest deflection calculated in this way.
2. Design each column so that, under the critical deflection and its design axial load, it develops at least the required moment. This is done using

$$M_{Sd}/M_{Sde} = e_{crit}/e_u[2 - (e_{crit}/e_u)] \quad (7.1)$$

where M_{Sd} is the design ultimate moment, M_{Sde} is the effective design ultimate moment, e_{crit} is the critical deflection, and e_u is the ultimate deflection of the column having an ultimate moment of M_{Sde}.

The above formula defines a column for which a deflection of e_{crit} corresponds to a moment of M_{Sd} by the application of assumption (d) above. The column section is then designed to resist the moment M_{Sde} under the action of the design axial load N_{Sd}. For the critical column, $M_{Sde} = M_{Sd}$.

For any given column, e_u may be written as a constant times K_2, hence the unknowns in the above equation are K_2 and M_{Sde}. The design charts in Figs 7.5 and 7.6 provide a relationship between these two variables, and hence an iterative solution is possible. The example below will clarify the method envisaged.

Example 7.2
The columns in the frame illustrated in Fig. 7.11 are to be designed to resist the design ultimate loads indicated. The rules in *clause 4.3.5.3.5* and *Fig. 4.27* have been used to calculate effective heights for the columns, and normal

Clause 4.3.5.3.5

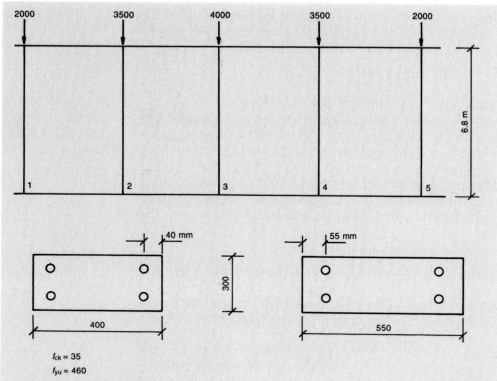

Fig. 7.11. *Example of sway frame*

Table 7.3. Example 7.2 data

Column	N_{sd}:kN	M_0:kN m	L_0:m	b:mm	h:mm	d'/h
1, 5	2000	100	8·84	300	400	0·1
2, 4	3500	75	8·16	300	400	0·1
3	4000	200	8·16	300	550	0·1

analytical procedures have been used to calculate the first-order moments. The results of these calculations are summarized in Table 7.3, together with the cross-sectional dimensions of the columns.

The next stage is to calculate the ultimate deflections of each column as if it was designed individually. This is done exactly as in Example 7.1, by calculating a basic deflection from *equations (4.72) and (4.69)* and then finding K_2 iteratively from Fig. 7.5. The minimum deflection found in this way is the critical deformation at which the structure will fail.

Having established the critical eccentricity, it is now necessary to establish effective ultimate moments for the sections that will ensure that each column will provide at least the design moment when the column is subjected to the critical eccentricity. For columns 2 and 4, since the critical deflection is equal to the ultimate deflection, the design moment equals the effective moment and so the column has to be designed for the moment in Table 7.4. Starting with column 3, the procedure for the other columns is as follows.

From equation (7.1)

$$279 \cdot 6/M_{Sde} = 19 \cdot 9/e_u[2 - (19 \cdot 9/e_u)]$$

For convenient use of the design charts, the moments should be divided by

SLENDER COLUMNS AND BEAMS

Table 7.4. Example 7.2 data

Column	$1/r \times 10^6$	e_2:mm	K_2	e_2:mm	Design moment
1, 5	9.66	$75.5K_2$	0.63	47.6	139.80
2, 4	9.66	$64.3K_2$	0.31	$19.9e_{crit}$	144.65
3	8.97	$59.7K_2$	0.60	36	279.60

$bh^2 f_{ck}$ and e_u should be expressed in terms of a constant multiplied by K_2. This constant can be found from Table 7.4. The equation now becomes

$$0.088/(M_{Sde}/bh^2 f_{ck}) = 19.9/59.7K_2[2-(19.9/59.7K_2)]$$

This can be rewritten as

$$(M_{Sde}/bh^2 f_{ck}) = 0.088/[1/3K_2(2-1/3K_2)]$$

By drawing a horizontal line across the design chart in Fig. 7.5 corresponding to the design axial load, it is possible to produce a relation between K_2 and $M/bh^2 f_{ck}$. The relation found in this way and the one given by equation (7.1) may be plotted on a graph. The point where the lines intersect gives the appropriate value of M_{Sde} (see Fig. 7.12). A similar operation can be carried out for columns 1 and 2. This leads to the following moments for which the columns should be designed.

Columns 1 and 5: 218 kN m
Columns 2 and 4: 145 kN m
Column 3: 285 kN m

Hence, column 3 has not required a major increase over the design moment in order to give compatibility of deformations at ultimate, but the edge columns, 1 and 5, have had to be designed for a very considerable increase in moment.

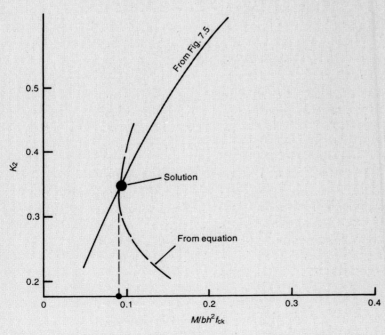

Fig. 7.12. *Relation of K_2 and M for sway example*

DESIGNERS' HANDBOOK TO EUROCODE 2

The effects of sway having been designed for the columns should now be checked individually as isolated braced columns. This is not done here, as the procedure is illustrated in Example 7.1.

7.3.5. Walls
Reinforced concrete walls may be treated by the simplified model column method in exactly the same way as columns, except that it is necessary only to consider the possibility of deformation about the minor axis (i.e. only deformation perpendicular to the plane of the wall need be considered). The approach adopted to the design of walls is that the first-order analysis for the vertical and lateral loads will give a distribution of axial load along the wall. Consideration of the interaction of the floors and beams framing into the walls will lead to transverse first-order moments. Sections of wall will then be designed for the maximum axial load per unit length of wall within the section considered, combined with any transverse moments and the appropriate accidental and second-order eccentricities.

7.3.6. Lateral buckling of slender beams
Where a beam is narrow compared with its span or depth, it may fail by lateral buckling of the form illustrated in Fig. 7.13. Beams where this is likely to be a

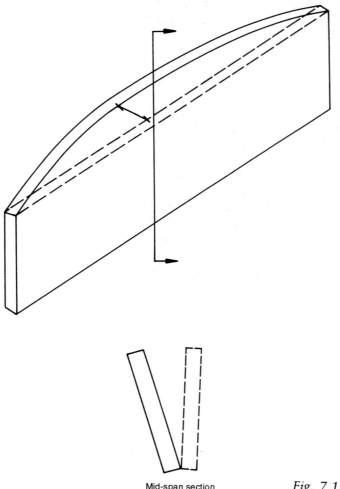

Mid-span section

Fig. 7.13. Buckling of slender beam

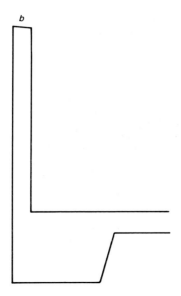

Fig. 7.14. Slender beam

problem are relatively rare, and hence a simple, conservative check is normally sufficient. EC2 includes such a check in *clause 4.3.5.7*. This states that the safety will be adequate without further check provided that

Clause 4.3.5.7

$$l_{ot} < 50b$$
$$h < 2 \cdot 5b$$

No indication is given in EC2 of how a further check should be formulated, should this be necessary. The draft CEB 1990 Model Code does, however, give a more detailed method of analysis in its clause 6.7.3.3.4. This method has similarities with the model column method in that it postulates an ultimate deflected shape and then ensures that the critical section can withstand the resulting internal actions. Reference should be made to the Model Code for details of the method.

A problem that occasionally occurs in practice with slender beams is where, for example, an edge beam is designed with a thin parapet cast monolithically with it, as shown in Fig. 7.14. Such a beam would normally be designed ignoring the effect of the upstand parapet, but rigorous interpretation of rules such as those in EC2 would imply that such a member cannot be used because of the slenderness of the parapet. It must, in such circumstances, be satisfactory to state that if the beam is adequately safe without the parapet, the addition of the parapet cannot make it less safe.

CHAPTER 8

Serviceability limit states

8.1. General

EC2 deals in some detail with three common serviceability limit states, i.e.

limitation of stresses *(clause 4.4.1)*
control of cracked *(clause 4.4.2)*
control of deflections *(clause 4.4.3* and *Appendix 4)*.

Clause 4.4.1
Clause 4.4.2
Clause 4.4.3
Appendix 4

Design for any limit state requires the definition of

(a) the appropriate loading and methods of analysis so that the design load effects can be established
(b) the design material properties to be assumed in the verification
(c) criteria defining the limit of satisfactory performance
(d) suitable methods by which performance may be predicted.

Of these, only (c) and (d) are found in *section 4.4*. Item (a) is found in *clauses 2.3.4 and 2.5.3.1*; item (b) is found in *Chapter 3*. All these matters are summarized in this chapter of the manual. In most cases it will not be necessary to carry out explicit calculations for the serviceability limit states, as simple 'deemed to satisfy' procedures are given in EC2 for dealing with all three of the limit states covered. This approach is acceptable because serviceability is intrinsically less critical than the ultimate limit states, and major calculation effort is not justified. For example, if the structure is wrongly designed and the strength is even 1% below the imposed loads, the structure collapses. If the crack width turns out to be 0·33 mm instead of 0·3 mm, nothing more serious than some grumbling from the owner and a cosmetic repair is likely to result.

Section 4.4
Clause 2.3.4
Clause 2.5.3.1
Chapter 3

8.1.1. Assessment of design action effects

Clause 2.3.4 defines three combinations of actions that may need to be considered when designing for a serviceability limit state

Clause 2.3.4

$$\text{rare combination: } G_{k,j} + (P) + Q_{k,1} + \psi_{0,i}Q_{k,i}$$

$$\text{frequent combination: } G_{k,j} + (P) + Q_{k,1} + \psi_{2,i}Q_{k,i}$$

$$\text{quasi-permanent combination: } G_{k,j} + (P) + \psi_{2,i}Q_{k,i}$$

It will be understood from these formulae that the partial factor on the permanent loads is always 1·0 for serviceability limit states. For buildings, *clause 2.3.4(7)* permits the following simplifications which may be used for the rare or frequent combinations

Clause 2.3.4(7)

(a) where there is only one variable action
$$G_{k,j} + (P) + Q_k$$
(b) where there are two or more variable actions
$$G_{k,j} + (P) + 0 \cdot 9 Q_{k,i}$$

According to *clause 2.5.3.2.1*, analysis should be elastic (without redistribution). This analysis may normally be based on the stiffness of the uncracked section; however, if it is suspected that cracking may have a significant unfavourable effect on the performance, then a more realistic analysis taking account of the cracking should be used. This possibility may be ignored for normal building structures.

The determination of the effects of prestress is covered in *clause 2.5.4*.

Clause 2.5.3.2.1

Clause 2.5.4

8.1.2. Material properties

The material properties that are normally significant in serviceability calculations are the modulus of elasticity of the reinforcement, and the modulus of elasticity, creep coefficient, shrinkage strain and tensile strength of the concrete. All this information is given in *Chapter 3*, but is summarized here for convenience.

Chapter 3

8.1.2.1. MODULUS OF ELASTICITY OF REINFORCEMENT. For ordinary reinforcement this may be taken as 200 kN/mm² *(clause 3.2.4.3)*.

For prestressing steel, 200 kN/mm² may be assumed for wires and bars and 190 kN/mm² for strand *(clause 3.3.4.4)*

Clause 3.2.4.3

Clause 3.3.4.4

8.1.2.2. MODULUS OF ELASTICITY OF CONCRETE. Figure 8.1 gives the value of the mean secant modulus of concrete E_{cm} as a function of the cylinder strength of the concrete at the time considered. The elastic modulus of concrete varies with other factors besides the strength (for example it varies with aggregate type), and, if an accurate prediction of serviceability conditions is required, it

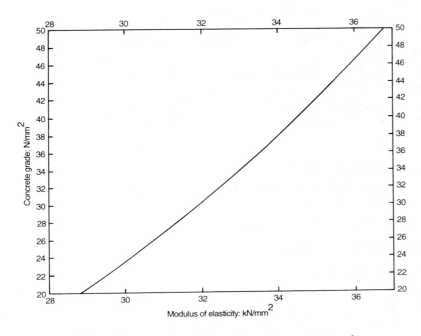

Fig. 8.1. Secant modulus of concrete as a function of concrete grade

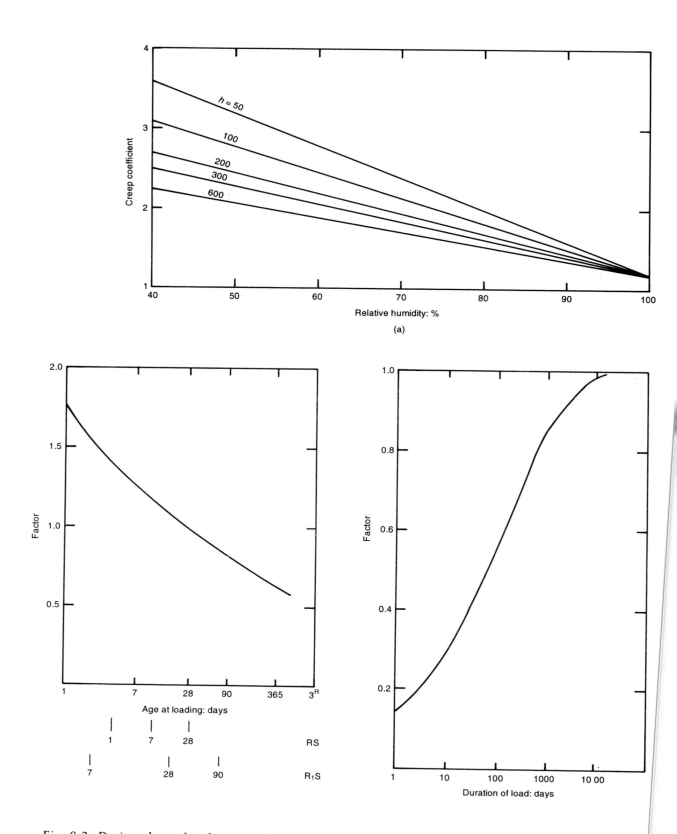

Fig. 8.2. Design charts for the approximate assessment of creep coefficients

SERVICEABILITY LIMIT STATES

will be necessary to establish the value of E_{cm} by tests on the type of concrete actually being used.

8.1.2.3 CREEP COEFFICIENT. Figure 8.2 gives values of the final creep coefficient for normal-weight concrete as a function of the notional thickness of the element, the relative humidity and the age at first loading. Fig. 8.2(a) gives the final creep coefficient as a function of relative humidity and notional thickness where initial loading takes place at 28 days. Fig. 8.2(b) gives factors by which this basic final creep factor should be multiplied to allow for ages at initial loading other than 28 days. Fig. 8.2(c) provides a correction factor for duration of loading, calculated for a section thickness of 150 mm: for thicker sections, it will be slightly conservative. All the parts of Fig. 8.2 have been derived from the equations given in *Appendix 1* assuming a characteristic concrete strength of 35 N/mm². *Appendix 1*

Appendix 1 gives further correction factors to allow for different types of cement and for temperature. Both these factors are taken into account by calculating a ficticious age at first loading, and have a significant effect only for ages below 28 days at first loading. *Clause 3.1.2.5.5.* gives a further factor which depends on the plastic consistency of the concrete. For stiff consistency S1, the creep coefficient should be multiplied by 0·7; for soft consistency S4, the creep should be multiplied by 1·2. It is interesting that the more rigorous approach of *Appendix 1* does not introduce any such factor. As with the elastic modulus, these values should be viewed as approximate and, if a high level of accuracy is necessary, creep coefficients should be obtained by test. The notional size of a section is, in fact, the thickness of a slab that would have the same creep characteristics. It is given by twice the cross-sectional area of the section divided by the section perimeter. It can easily be shown that the notional size of a slab of infinite width is equal to its actual overall depth.

Clause 3.1.2.5.5

Appendix 1

8.1.2.4. FREE SHRINKAGE STRAIN. The free shrinkage strain can be obtained from Fig. 8.3. Fig. 8.3(a) gives values for the final free shrinkage strain; Fig. 8.3(b) gives the fraction of the free shrinkage that can be expected to have occurred after specified lengths of time. *Clause 3.1.2.5.5* suggests the same multipliers to allow for consistency as are suggested for creep. These are

Clause 3.1.2.5.5

- stiff consistency (S1): multiply shrinkage by 0·7
- normal consistency (S2 and S3): 1·0
- soft consistency (S4): 1·2

Again, the more rigorous approach set out in *Appendix 1* does not introduce any correction for consistency.

Appendix 1

8.1.2.5. TENSILE STRENGTH OF CONCRETE. Figure 8.4 gives the main tensile strength and the upper and lower characteristic tensile strengths as a function of the characteristic compressive strength of the concrete. The values are derived using *equations (3.2)–(3.4)* in *clause 3.1.2.3*.

Clause 3.1.2.3

8.2. Limitation of stresses under serviceability conditions

8.2.1. General

The question of the possible limitation of stresses under serviceability loads was the subject of a great deal of discussion during the drafting of EC2. The basic reason for differences of opinion was the different ways in which design methods had developed in different countries, and a natural desire in each country that

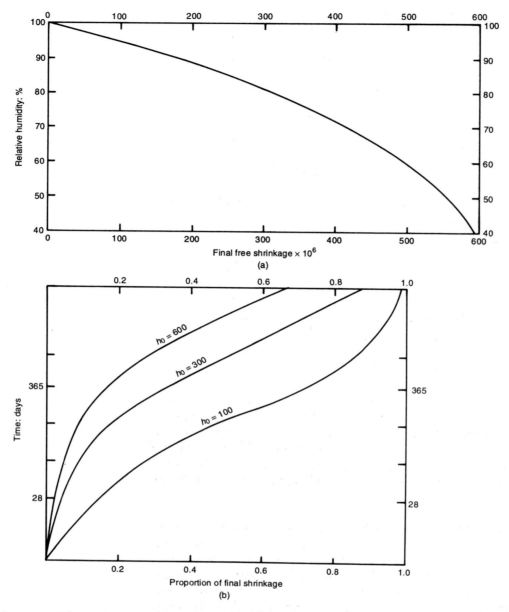

Fig. 8.3. Charts for the assessment of free shrinkage strain

the new and untried Eurocode should not force design too far from existing experience in that country. Limitations on the compressive stress in the concrete were particularly contentious. In many countries, section design for both prestressed and reinforced concrete has been carried out on the basis of ultimate strength methods, and no explicit check has been required on the concrete stresses under service loads. These countries were determined that no such checks should be introduced, since these could be shown to have a significant effect on the economy of some types of member, notably columns. Other countries have not yet adopted ultimate load methods, and still base their design on an elastic analysis of sections combined with a stress limit. While these countries were prepared to accept ultimate strength methods for section design, they wanted to retain elastic design as an additional requirement to ensure that member sizes, particularly columns, did not become much smaller or more lightly reinforced than those arising

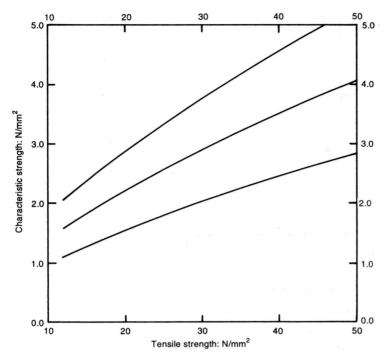

Fig. 8.4. Tensile strength of concrete

from their current design methods. The check on the compressive stress in the concrete at the serviceability limit state provided them with this safeguard.

In order to develop a rational draft for this section of EC2 which would have some hope of general acceptance, an attempt was made to establish what rational basis existed for a limit on stresses under serviceability conditions. Two types of limit need to be considered: tensile stress limits and compressive stress limits. These are now considered.

8.2.1.1. LIMITS TO COMPRESSIVE STRESS. Two reasons are usually put forward for limiting compressive stresses in concrete: to avoid the formation of micro-cracks in the concrete, which might reduce durability, and to avoid excessive creep. There is some logic is both reasons. It is commonly accepted that micro-cracking will start to develop in concrete when the compressive stress exceeds about 70% of the compressive strength. However, none of the established publications on durability suggest any relationship between stress level, micro-cracking and durability problems (for example see Ref. 11). Furthermore, were such a relationship seriously believed to be a problem, one would expect the stress limit to be a function of the aggressivity of the environment. Nevertheless, EC2 takes a conservative line on this issue and suggests that in aggressive environments (exposure classes 3 and 4) it may be reasonable to limit the stress to $0 \cdot 6 f_{ck}$ if other measures, such as an increase in the cover to any compressive reinforcement or the provision of confinement to the compressed concrete by transverse reinforcement, are taken.

The design methods used for checking serviceability assume that creep will be proportional to stress. This assumption is increasingly invalid, as the stress increases above about 50% of the compressive strength. It could therefore be logical to impose a limit on compressive stress under the quasi-permanent combination of loads, where a higher level of creep than might be predicted could seriously impair performance.

It will be seen that there is no case for a blanket limitation to compressive stresses, but that in particular cases it may be prudent to apply such limits. *Clause 4.4.1.1(1) –(3)* apply this philosophy.

Clause 4.4.1.1(1)
Clause 4.4.1.1(2)
Clause 4.4.1.1(3)

8.2.1.2. LIMITATION OF TENSILE STRESSES.
Limits may be defined for the tensile stresses in the concrete or in the reinforcement. In the case of reinforcement, it seems reasonable to ensure that inelastic deformations of the steel are avoided under service loads. Such deformations (*a*) would invalidate any calculations of cracking or deflections which assume that the reinforcement behaves elastically and (*b*) could result in excessively large cracks. This could be particularly critical in areas subjected to frequent variations in loading. The limits suggested in *clause 4.4.1.1(7)* are aimed at ensuring that inelastic deformations do not occur, and include some margin of safety for this.

Clause 4.4.1.1(7)

Prestressing steels are much more sensitive to corrosion damage than normal reinforcing bars. This is partly due to differences in metallurgy, but mostly due to their smaller diameter and the high level of stress at which they normally operate. A prestressing tendon may frequently operate for its whole life at a stress of up to 70% of its tensile strength over its whole length. A reinforcing bar is likely to be stressed above about 50% of its yield strength only for insignificant periods, and then only in a very limited area close to sections of maximum moment. It is therefore necessary to take much more stringent precautions against corrosion with prestressed members than with reinforced members. In any parts of a member that could be exposed to salt, it is necessary to ensure that no direct paths exist that could give the salt direct access to the surface of the tendon. Cracks could provide such a path, and so cracks that could penetrate to the tendons should be avoided in salty environments. This objective can conveniently be achieved by ensuring that no tension develops in the concrete surrounding the tendons. *Clause 4.4.2.1(7)* provides an element of extra safety by requiring that, in environments where this precaution is necessary, the tendon is surrounded by at least 25 mm of uncracked concrete.

Clause 4.4.2.1(7)

8.2.2. Procedure for stress checks
In general, stress checks may be avoided for reinforced concrete, since *clause 4.1.1.2(2)* states that the stress checks may be deemed to have been satisfied provided that four conditions have been met, i.e. that

Clause 4.1.1.2(2)

(*a*) the design for the ultimate limit state has been carried out in accordance with *section 4.3*

Section 4.3

(*b*) the minimum reinforcement provisions of *clause 4.4.2.2* have been satisfied

Clause 4.4.2.2

(*c*) detailing is carried out according to *Chapter 5*

Chapter 5

(*d*) not more then 30% of redistribution has been carried out.

Thus, provided design is carried out in the normal way envisaged by EC2, there will be no need to carry out stress checks unless there are specific reasons why they should be necessary for the design in question.

If stresses are to be checked, the calculation is performed on the following basis.

(*a*) Plane sections remain plane.
(*b*) Reinforcement and concrete in tension are assumed to be elastic.
(*c*) Concrete is assumed to be elastic up to its tensile strength f_{ctm}. If the stress has exceeded f_{ctm}, the section is assumed to be cracked and the concrete in tension is ignored.

(d) Creep may generally be taken into account by assuming that the ratio of the elastic modulus for steel to that for concrete (modular ratio) is 15. A lower value, based on the actual elastic modulus of the concrete, may be used where less than 50% of the stresses arise from quasi-permanent loads. A more accurate assessment of the effects of creep may, of course, be used if desired.

EC2 states that shrinkage and temperature stresses should be taken into account where they are likely to be significant. This will not usually be necessary for normal reinforced and prestressed members in buildings, but appropriate methods are considered in section 8.2.2.1 below.

In general, stress checks can be carried out using standard elastic formulae. Thus, for beams subjected only to flexure, stresses may be calculated from

$$\text{stress} = My/I \qquad (8.1)$$

where M is the applied moment, y is the distance from the neutral axis to the point considered, and I is the second moment of area of the section (this may be based on either a cracked or an uncracked section, as appropriate).

Similarly, for uncracked reinforced members subject to moments and axial load or prestressed beams, the stresses may be calculated from the standard relation

$$\text{stress} = My/I + N/A_c \qquad (8.2)$$

where N is the applied axial force and A_c is the area of the total concrete section.

Tables 8.1–8.4 give values for the neutral axis depth and second moments of areas of cracked and uncracked rectangular and flanged sections. For cracked sections, only singly reinforced members are considered.

Table 8.1. Neutral axis depths and moments of inertia for flanged beams: $hf/d = 0.2$

| | \multicolumn{10}{c}{b_r/b} |
| | 1 | | 0.5 | | 0.4 | | 0.3 | | 0.2 | |
α_{ep}	x/d	I/bd^3	x/d	I/bd^3	x/d	I/bd^3	x/d	I/bd^3	x/d	I/bd^3
0.020	0.181	0.015								
0.030	0.217	0.022	0.217	0.022	0.217	0.022	0.217	0.022	0.217	0.022
0.040	0.246	0.028	0.248	0.028	0.248	0.028	0.249	0.028	0.249	0.028
0.050	0.270	0.033	0.274	0.033	0.275	0.033	0.276	0.033	0.278	0.033
0.060	0.292	0.038	0.298	0.038	0.300	0.038	0.302	0.038	0.304	0.038
0.070	0.311	0.043	0.320	0.043	0.322	0.043	0.325	0.043	0.327	0.043
0.080	0.328	0.048	0.340	0.047	0.343	0.047	0.346	0.047	0.349	0.047
0.090	0.344	0.052	0.358	0.052	0.361	0.052	0.365	0.051	0.369	0.051
0.100	0.358	0.057	0.375	0.056	0.379	0.056	0.383	0.055	0.388	0.055
0.110	0.372	0.061	0.390	0.060	0.395	0.059	0.400	0.059	0.406	0.059
0.120	0.384	0.064	0.405	0.063	0.410	0.063	0.416	0.063	0.422	0.062
0.130	0.396	0.068	0.418	0.067	0.424	0.066	0.430	0.066	0.437	0.065
0.140	0.407	0.072	0.431	0.070	0.437	0.070	0.444	0.069	0.452	0.069
0.150	0.418	0.075	0.443	0.073	0.450	0.073	0.457	0.072	0.466	0.071
0.160	0.428	0.078	0.455	0.076	0.462	0.076	0.470	0.075	0.478	0.074
0.170	0.437	0.082	0.466	0.079	0.473	0.078	0.481	0.078	0.491	0.077
0.180	0.446	0.085	0.476	0.082	0.484	0.081	0.493	0.080	0.502	0.079
0.190	0.455	0.088	0.486	0.085	0.494	0.084	0.503	0.083	0.513	0.082
0.200	0.463	0.091	0.495	0.087	0.504	0.086	0.513	0.085	0.524	0.084
0.210	0.471	0.094	0.504	0.090	0.513	0.089	0.523	0.088	0.534	0.086
0.220	0.479	0.096	0.513	0.092	0.522	0.091	0.532	0.090	0.543	0.089
0.230	0.486	0.099	0.521	0.094	0.531	0.093	0.541	0.092	0.552	0.091
0.240	0.493	0.102	0.529	0.097	0.539	0.095	0.549	0.094	0.561	0.093

Table 8.2. Neutral axis depths and moments of inertia for flanged beams: $h_f/d = 0.3$

			\multicolumn{8}{c}{b_r/b}					
	\multicolumn{2}{c}{1}	\multicolumn{2}{c}{0.5}	\multicolumn{2}{c}{0.4}	\multicolumn{2}{c}{0.3}	\multicolumn{2}{c}{0.2}
p	x/d	I/bd^3	x/d	I/bd^3	x/d	I/bd^3	x/d	I/bd^3	x/d	I/bd^3
0.020	0.181	0.015								
0.030	0.217	0.022								
0.040	0.246	0.028								
0.050	0.270	0.033								
0.060	0.292	0.038								
0.070	0.311	0.043	0.311	0.043	0.311	0.043	0.311	0.043	0.311	0.043
0.080	0.328	0.048	0.328	0.048	0.329	0.048	0.329	0.048	0.329	0.048
0.090	0.344	0.052	0.345	0.052	0.345	0.052	0.345	0.052	0.346	0.052
0.100	0.358	0.057	0.360	0.056	0.361	0.056	0.361	0.056	0.362	0.056
0.110	0.372	0.061	0.375	0.060	0.375	0.060	0.376	0.060	0.377	0.060
0.120	0.384	0.064	0.388	0.064	0.389	0.064	0.390	0.064	0.391	0.064
0.130	0.396	0.068	0.401	0.068	0.402	0.068	0.403	0.068	0.404	0.068
0.140	0.407	0.072	0.413	0.071	0.414	0.071	0.416	0.071	0.417	0.071
0.150	0.418	0.075	0.425	0.075	0.426	0.075	0.428	0.075	0.430	0.075
0.160	0.428	0.078	0.436	0.078	0.437	0.078	0.439	0.078	0.441	0.078
0.170	0.437	0.082	0.446	0.081	0.448	0.081	0.450	0.081	0.452	0.081
0.180	0.446	0.085	0.456	0.084	0.458	0.084	0.461	0.084	0.463	0.084
0.190	0.455	0.088	0.466	0.087	0.468	0.087	0.471	0.087	0.473	0.087
0.200	0.463	0.091	0.475	0.090	0.477	0.090	0.480	0.090	0.483	0.089
0.210	0.471	0.094	0.483	0.093	0.486	0.092	0.489	0.092	0.493	0.092
0.220	0.479	0.096	0.492	0.095	0.495	0.095	0.498	0.095	0.502	0.095
0.230	0.486	0.099	0.500	0.098	0.503	0.098	0.507	0.097	0.511	0.097
0.240	0.493	0.102	0.508	0.100	0.511	0.100	0.515	0.100	0.519	0.099

Table 8.3. Neutral axis depths and moments of inertia for flanged beams: $h_f/d = 0.4$

	\multicolumn{2}{c}{1}	\multicolumn{2}{c}{0.5}	\multicolumn{2}{c}{0.4}	\multicolumn{2}{c}{0.3}	\multicolumn{2}{c}{0.2}
p	x/d	I/bd^3	x/d	I/bd^3	x/d	I/bd^3	x/d	I/bd^3	x/d	I/bd^3
0.020	0.181	0.015								
0.030	0.217	0.022								
0.040	0.246	0.028								
0.050	0.270	0.033								
0.060	0.292	0.038								
0.070	0.311	0.043								
0.080	0.328	0.048								
0.090	0.344	0.052								
0.100	0.358	0.057								
0.110	0.372	0.061								
0.120	0.384	0.064								
0.130	0.396	0.068								
0.140	0.407	0.072	0.407	0.072	0.407	0.072	0.407	0.072	0.407	0.072
0.150	0.418	0.075	0.418	0.075	0.418	0.075	0.418	0.075	0.418	0.075
0.160	0.428	0.078	0.428	0.078	0.428	0.078	0.428	0.078	0.428	0.078
0.170	0.437	0.082	0.438	0.082	0.438	0.082	0.438	0.082	0.438	0.082
0.180	0.446	0.085	0.447	0.085	0.447	0.085	0.448	0.085	0.448	0.085
0.190	0.455	0.088	0.456	0.088	0.457	0.088	0.457	0.088	0.457	0.088
0.200	0.463	0.091	0.465	0.091	0.465	0.091	0.466	0.091	0.466	0.091
0.210	0.471	0.094	0.473	0.094	0.474	0.094	0.474	0.094	0.474	0.093
0.220	0.479	0.096	0.481	0.096	0.482	0.096	0.482	0.096	0.483	0.096
0.230	0.486	0.099	0.489	0.099	0.490	0.099	0.490	0.099	0.491	0.099
0.240	0.493	0.102	0.496	0.101	0.497	0.101	0.498	0.101	0.498	0.101

Table 8.4. Properties of uncracked sections

	\multicolumn{8}{c|}{h_f/h}							
	1		0.4		0.3		0.2	
b_r/b	x/h	I/bh^3	x/h	I/bh^3	x/h	I/bh^3	x/h	I/bh^3
0.100			0.265	0.020	0.245	0.019	0.243	0.019
0.150			0.292	0.026	0.280	0.026	0.288	0.026
0.200			0.315	0.032	0.309	0.032	0.322	0.031
0.250			0.336	0.037	0.334	0.037	0.350	0.036
0.300	0.500	0.083	0.355	0.042	0.356	0.042	0.373	0.041
0.350			0.372	0.046	0.375	0.046	0.392	0.045
0.400			0.388	0.050	0.391	0.050	0.408	0.049
0.450			0.401	0.054	0.406	0.054	0.421	0.052
0.500			0.414	0.057	0.419	0.057	0.433	0.055

Figures 8.5 and 8.6 give values for the cracked neutral axis depths and second moments of area of doubly reinforced rectangular sections. These have been taken from Ref. 12.

The calculation of stresses in a cracked member subject to axial forces and moments is less simple, since the second moment of area of the cracked section is a function of the axial force. In such cases it is generally easier to write out the equations of equilibrium and solve these iteratively. An approximate value of the steel stress in such sections can, however, be calculated by use of the design chart in Fig. 8.7, as follows.

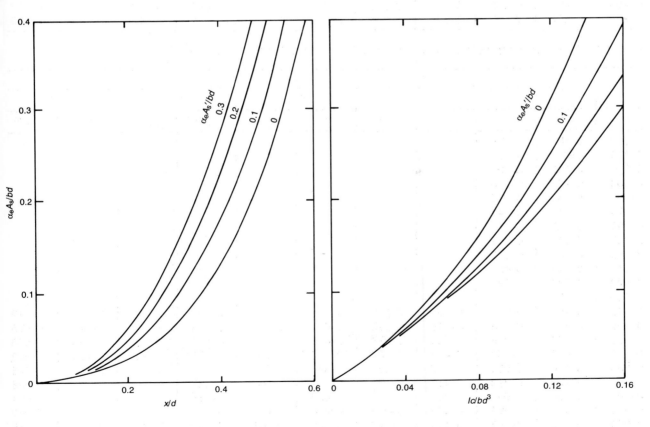

Fig. 8.5. Neutral axis depths calculated on the basis of a cracked section

Fig. 8.6. Second moments of area of rectangular sections calculated on the basis of a cracked section

DESIGNERS' HANDBOOK TO EUROCODE 2

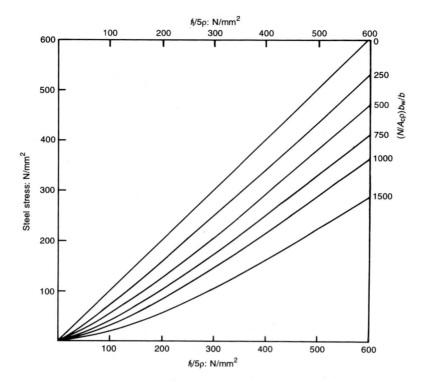

Fig. 8.7. Design chart for the approximate calculation of the stress in reinforcement in a prestressed beam

(a) Calculate the average axial stress in the section by dividing the longitudinal force by the area of the concrete section. This is then divided by the reinforcement ratio $A_s/b_w d$ and multiplied by the factor b_w/b to give the longitudinal force factor $Nb_w/(A_c b)$.

(b) Calculate the 'notional tensile stress' f_t at the tension face of the section. This is calculated on the basis of an uncracked section, even though the section is cracked, and is then used to calculate the factor $f_t/5$.

(c) The above two factors can then be used in Fig. 8.7 to give the steel stress in any unprestressed reinforcement in a prestressed beam or the change in stress in the prestressing tendons from the condition where the stress in the concrete immediately surrounding the steel is zero.

Example 8.1

Calculate the stresses in a T-beam with the following dimensions when subjected to a moment of 100 kN m: overall breadth, 600 mm; overall depth, 500 mm; effective depth, 450 mm; flange depth, 150 mm; rib breadth, 200 mm; steel area, 2455 mm². The concrete grade is C30/37.

From Fig. 8.4, the mean tensile strength of the concrete is 2·8 N/mm². From Table 8.4, the neutral axis depth and second moment of area of the uncracked section can be found to be

$$x_I = 0 \cdot 367 \times 500 = 184 \text{ mm}$$

$$I_I = 0 \cdot 045 \times 600 \times 500^3 = 3375 \times 10^6 \text{ mm}^4$$

The tensile stress at the bottom of the section, assuming an uncracked section, is thus given by

$$f_t = 100\,(500-184)/3375 = 9 \cdot 36 \text{ N/mm}^2$$

This is greater than the tensile strength of the concrete, so the section should be assumed to be cracked.

The neutral axis depth and second moment of area of the cracked section can be obtained by interpolation from Tables 8.2 and 8.3.

The reinforcement ratio multiplied by the modular ratio is given by

$$p = 2455/600/450 \times 15 = 0 \cdot 136$$

From the tables

$$x_{II} = 0 \cdot 41 \times 450 = 185 \text{ mm}$$

$$I_{II} = 0 \cdot 07 \times 600 \times 450^3 = 3827 \times 10^6 \text{ mm}^4$$

The stresses can now be calculated as

$$\text{compressive stress in concrete} = 100 \times 185/3827 = 4 \cdot 83 \text{ N/mm}^2$$

$$\text{tensile stress in steel} = 100 \times (450-185) \times 15/3827 = 104 \text{ N/mm}^2$$

Both these stress are well below the limits given in EC2.

8.2.2.1. STRESS CHECKS FOR REINFORCED CONCRETE COLUMNS.

The basic approach for checking the stresses in columns is the same as for beams, and, for uncracked members, the formulae given in section 8.2.2 above for combined axial force and bending may be used. For cracked sections, the algebra becomes more involved, and the design charts given in Figs 8.8–8.31 may be used. The first set of charts (Figs 8.8–8.19) gives the reinforcement areas necessary to meet the limit of $0 \cdot 6 f_{ck}$; the second set (Figs 8.20–8.31) gives the reinforcement areas necessary to meet the limit of $0 \cdot 45 f_{ck}$.

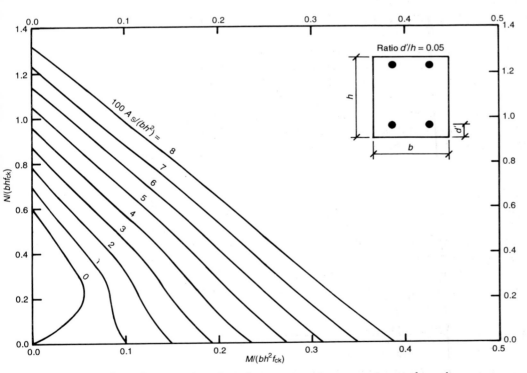

Fig. 8.8. *Design chart for ensuring that the compressive stress in reinforced concrete columns does not exceed* $0 \cdot 6 f_{ck}$

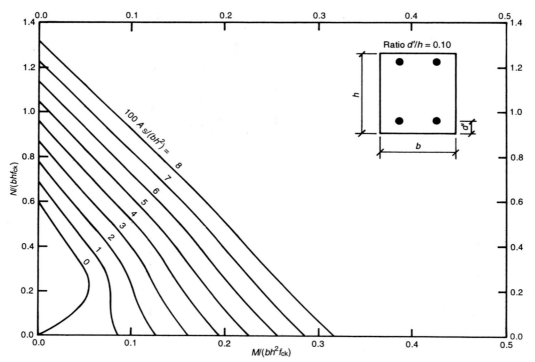

Fig. 8.9. Design chart for ensuring that the compressive stress in reinforced concrete columns does not exceed $0 \cdot 6 f_{ck}$

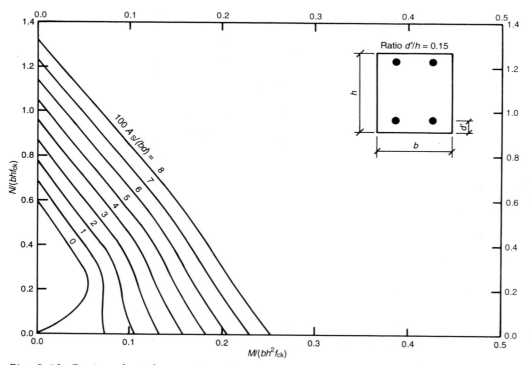

Fig. 8.10. Design chart for ensuring that the compressive stress in reinforced concrete columns does not exceed $0 \cdot 6 f_{ck}$

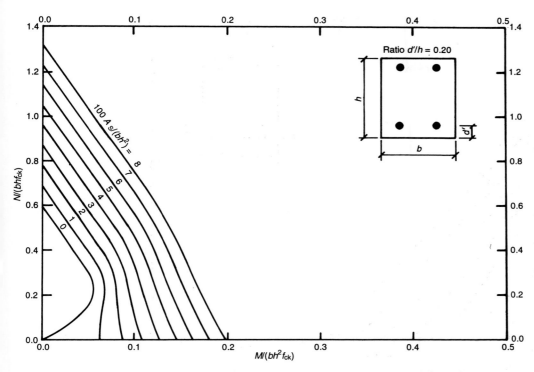

Fig. 8.11. Design chart for ensuring that the compressive stress in reinforced concrete columns does not exceed $0 \cdot 6f_{ck}$

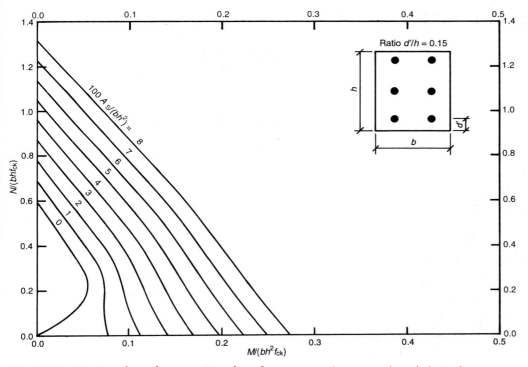

Fig. 8.12. Design chart for ensuring that the compressive stress in reinforced concrete columns does not exceed $0 \cdot 6f_{ck}$

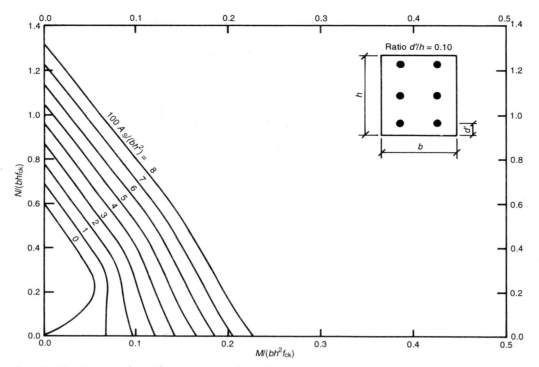

Fig. 8.13. Design chart for ensuring that the compressive stress in reinforced concrete columns does not exceed $0 \cdot 6 f_{ck}$

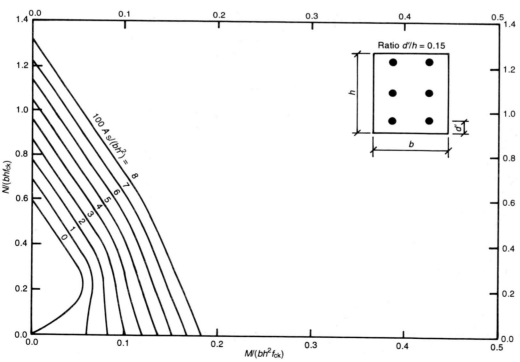

Fig. 8.14. Design chart for ensuring that the compressive stress in reinforced concrete columns does not exceed $0 \cdot 6 f_{ck}$

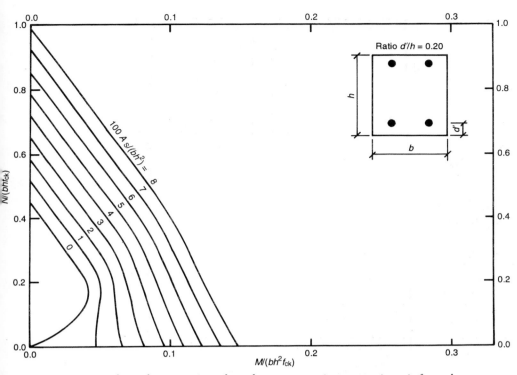

Fig. 8.23. Design chart for ensuring that the compressive stress in reinforced concrete columns does not exceed $0 \cdot 6 f_{ck}$

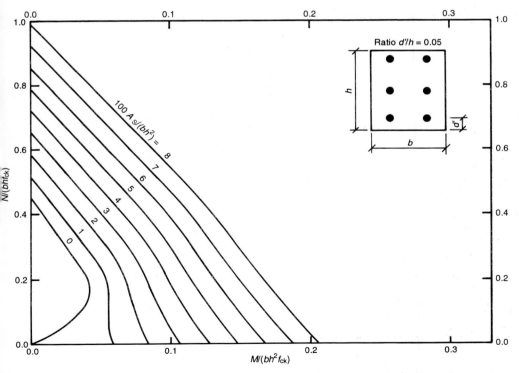

Fig. 8.24. Design chart for ensuring that the compressive stress in reinforced concrete columns does not exceed $0 \cdot 6 f_{ck}$

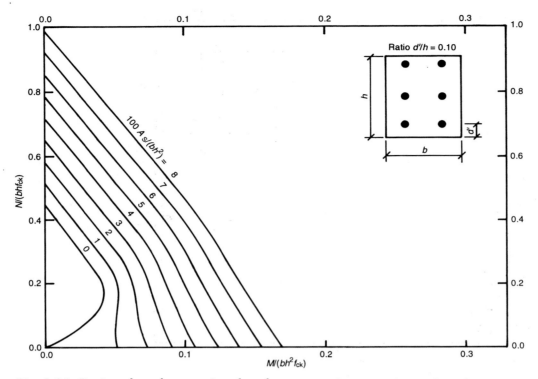

Fig. 8.25. Design chart for ensuring that the compressive stress in reinforced concrete columns does not exceed $0 \cdot 6 f_{ck}$

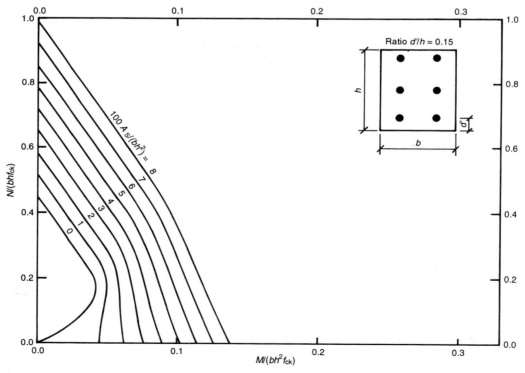

Fig. 8.26. Design chart for ensuring that the compressive stress in reinforced concrete columns does not exceed $0 \cdot 6 f_{ck}$

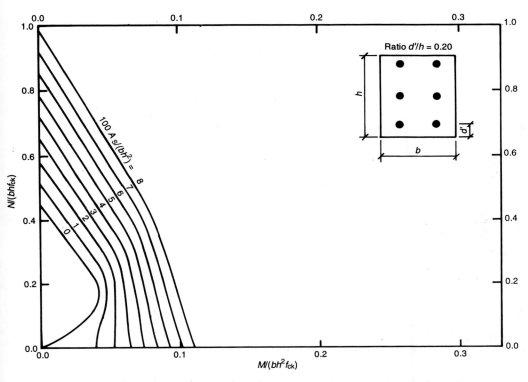

Fig. 8.27. Design chart for ensuring that the compressive stress in reinforced concrete columns does not exceed $0·6f_{ck}$

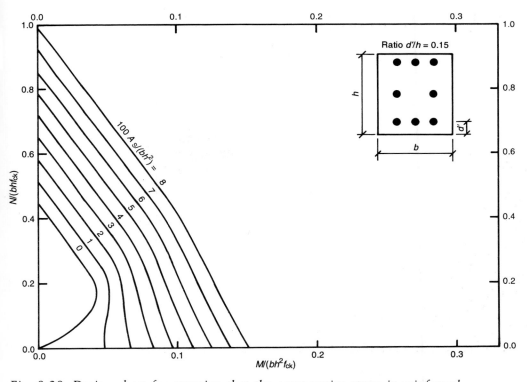

Fig. 8.28. Design chart for ensuring that the compressive stress in reinforced concrete columns does not exceed $0·6f_{ck}$

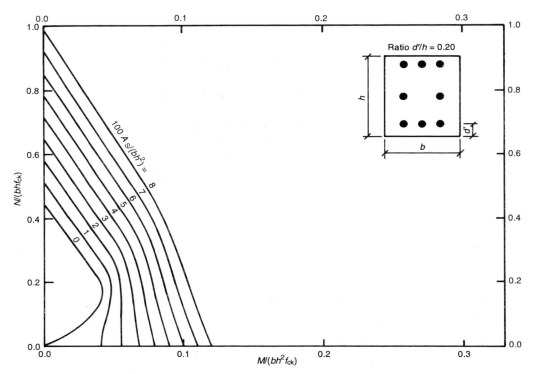

Fig. 8.29. Design chart for ensuring that the compressive stress in reinforced concrete columns does not exceed $0\cdot 6f_{ck}$

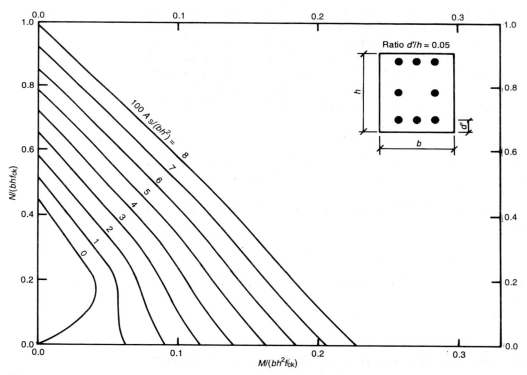

Fig. 8.30. Design chart for ensuring that the compressive stress in reinforced concrete columns does not exceed $0\cdot 6f_{ck}$

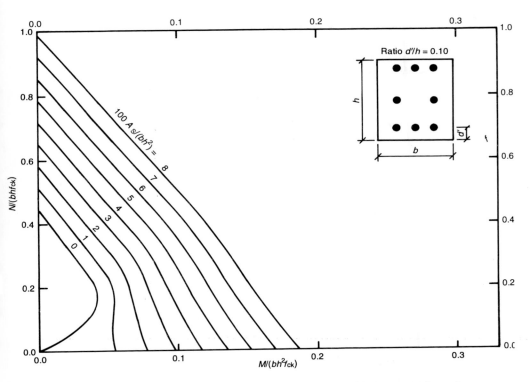

Fig. 8.31. *Design chart for ensuring that the compressive stress in reinforced concrete columns does not exceed* $0 \cdot 6f_{ck}$

8.2.2.2. STRESSES DUE TO TEMPERATURE CHANGES. A problem that occasionally has to be dealt with in deeper members is the estimation of the stresses induced in a member due to non-uniform temperature distributions. For uncracked and unrestrained members, well-established methods of analysis exist. Under the influence of non-uniform and non-linear temperature changes, the member will change in overall length and will also bend. The overall length change will be proportional to the average temperature change. For any general section subject to any general distribution of temperature, the average shortening will be given by

$$\epsilon_{av} = \alpha/A \int_0^h b_y t_y dy$$

where A is the total area of the section, h is the overall depth of the section, b_y is the breadth of the section at depth y, t_y is the temperature change at depth y, and α is the coefficient of expansion.

The curvature can be calculated by taking moments about the section centroid. This gives the relation

$$1/r = \alpha/I \int_0^h b_y t_y (y - Y) dy$$

where Y is the depth to the centroid of the section and I is the second moment of area of the section.

The actual strain at level y is now given by

$$\epsilon_{ay} = \epsilon_{av} + 1/r(y - Y)$$

The stress at any level can now be obtained from the difference between the actual strain and the free temperature movement at that level. Thus

$$\sigma_y = E_c(\alpha t_y - \epsilon_{ay})$$

While these equations may look complicated, they can in fact be solved quite easily by numerical methods. A very simple approach is to use a spreadsheet program.

The above method has been developed assuming that the member considered is unrestrained. The effects of restraint can be introduced into the analysis by introducing springs at the member ends that will resist axial and rotational movement. An iterative calculation is then required to develop a situation giving equilibrium between the spring forces and the forces in the member.

A similar approach can be adopted to deal with cracked sections subjected to temperature change. The method becomes highly iterative and is beyond the scope of this manual. In general, cracking, by reducing the stiffness of the member, will lead to a major relaxation in the stresses induced. It will therefore generally be satisfactory to ignore temperature-induced stresses in cracked members.

8.3. Control of cracking

8.3.1. General
This section covers the material in *clause 4.4.2*.

Clause 4.4.2

8.3.1.1. CRACK WIDTH LIMITS.   There are many reasons for wishing to limit the widths of cracks to a relatively small value. The most commonly cited reasons include the following.

(a) To avoid possible corrosion damage to the reinforcement due to deleterious substances penetrating to the reinforcement down the cracks.
(b) To avoid, or limit, leakage through cracks. This is commonly a critical design consideration in water-retaining structures.
(c) To avoid an unsightly appearance.

All three of these reasons have been researched to some degree but no clear definition of permissible crack width has emerged. The results of these studies are now summarized briefly.

8.3.1.1.1. Cracking and corrosion.   This is the most extensively researched area. Summaries of the findings have been published by a number of authors (see for example Refs 13 and 14). The development of corrosion is a two-phase process. In fresh concrete the reinforcement is protected from corrosion by the alkaline nature of the concrete. This protection can be destroyed by two mechanisms: carbonation of the concrete to the surface of the reinforcement or ingress of chlorides. Cracks will lead to a local acceleration of both processes by permitting more rapid ingress of either carbon dioxide of chlorides to the surface of the reinforcement. Once the protection provided by the concrete has been destroyed, corrosion can start if the environmental conditions are right. The period from construction to the initiation of corrosion is usually referred to as the 'initiation phase'; the period after the initiation of corrosion is termed the 'active phase'. The length of the initiation phase is likely to be influenced by crack width. However, this period is likely to be short at a crack, and some corrosion can usually be found on the bar surface where a crack reaches a bar after as little as two years, even with very small cracks. It is found, however, that this initial corrosion does not develop in cases where the cracks are small or where the bars intersect the cracks. The corrosion products possibly block the cracks and inhibit further corrosion. A more serious situation exists where a crack runs along the line of a bar. There is limited evidence to suggest that, in this case, sustained corrosion may develop in salty environments where the cracks exceed about 0·3 mm.

Although less research has been done on the relation of cracking and corrosion in prestressed concrete members, it is generally believed that the risks posed by cracks are greater, and therefore more stringent criteria should be imposed. For this reason, EC2 does not permit cracks to penetrate to the prestressing tendons where the member is exposed to aggressive environments.

8.3.1.1.2. Leakage. Only very limited research has so far been carried out into this problem, and this has not led to any agreed basis for crack width limits. Practical experience has suggested that cracks less than 0·2 mm wide that pass right through a section will leak somewhat initially, but will quickly seal themselves. This problem is not specifically considered in Part 1 of EC2, as liquid-retaining structures will be covered in Part 4.

8.3.1.1.3. Appearance. Limited studies suggest that noticeable cracks in structural members cause concern to the occupants of structures, and it is therefore advisable to keep cracks down to a width that will not generally be noticed by a casual observer. On a smoothly finished concrete surface, it appears that cracks are unlikely to lead to complaint if the maximum width is kept below 0·3 mm. Clearly, larger widths could be used on rougher forms of surface.

The general considerations outlined above led to the limits on crack width set out in EC2. However, the information on which the criteria were based is far from unambiguous.

8.3.1.2. CAUSES OF CRACKING. There are many possible causes of cracking, only a few of which lead to cracks that can be controlled by measures taken during the design. The following are the more common causes.

(a) Plastic shrinkage or plastic settlement: these are phenomena that occur within the first few hours after casting while the concrete is still in a plastic state. The likelihood of cracks being caused by these phenomena depends on the bleeding rate of the mix and the evaporation rate. The resulting cracks may be large; up to 2 mm is not uncommon.
(b) Corrosion: rust occupies a greater volume than the metal from which it is formed. Its formation therefore causes internal pressures to build up around the bar surface, which will lead to the formation of cracks running along the line of the corroding bars and, eventually, to spalling of the concrete cover.
(c) Expansive chemical reactions within the concrete: expansive reactions occurring at the concrete surface tend to lead to scaling of the concrete rather than cracking. However, some reactions, such as alkali–silica reaction, occur within the body of the concrete and can lead to large surface cracks.
(d) Restrained deformations such as shrinkage or temperature movements.
(e) Loading.

Of the above items, only (d) and (e) can be treated by the designer. These are probably the two least serious causes of cracking.

8.3.2. Principles of the cracking phenomena
The development of cracking is now described, for simplicity, in terms of a reinforced concrete member subjected to pure axial tension.

8.3.2.1. CRACKING CAUSED BY LOADING. If a continuously increasing tension is applied to a tension member, the first crack will form when the tensile strength of the weakest section in the member is exceeded. The formation of this crack leads to a local redistribution of stresses within the section. At the crack, all the tensile force will be transferred to the reinforcement and the stress in the concrete immediately adjacent to the crack must clearly be zero. With increasing distance from the crack, force is transferred by bond from the reinforcement to the concrete until, at some distance S_0 from the crack, the stress distribution within the section is unchanged from what it was before the crack formed. This local redistribution of forces in the region of the crack is accompanied by an extension of the member. This extension, plus a minor shortening of the concrete which has been relieved of the tensile stress it was supporting, is accommodated in the crack. The crack thus opens up to a finite width immediately on its formation. The formation of the crack and the resulting extension of the member also reduce the stiffness of the member. As further load is applied, a second crack

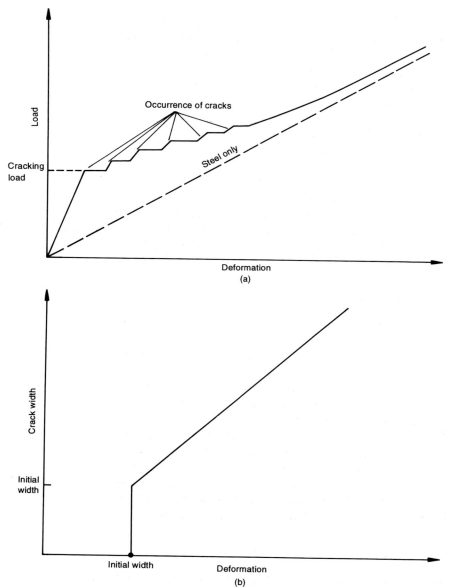

Fig. 8.32. (a) Load–deformation response of a member subjected to steadily increasing load; (b) crack width–deformation response in a load-controlled test

will form at the next weakest section, although it will not form within S_0 of the first crack since the stresses within this region will have been reduced by the formation of the first crack. Further increases in loading will lead to the formation of further cracks until, eventually, no area of the member surface remains that is not within S_0 of a previously formed crack. The formation of each crack will lead to a reduction in the member stiffness. After all the cracks have formed, further loading will result in a widening of the existing cracks but no new cracks. Stresses in the concrete will be relieved by limited bond slip near the crack faces and by the formation of internal cracks. This process leads to further reductions in stiffness but, clearly, the stiffness cannot reduce to below that of the bare reinforcement. Fig. 8.32 illustrates the behaviour described above. The stepped nature of the load–deformation response is not normally discernible in tests, and the response is usually shown as a smooth curve.

8.3.2.2. CRACKING DUE TO IMPOSED DEFORMATIONS. If the member is subjected to a steadily increasing strain rather than a steadily increasing load, the response illustrated in Fig. 8.32 is radically changed. When a crack forms, the reduction in stiffness resulting from the formation of the crack leads to a reduction in the tensile force supported by the member. This, in turn, leads to a reduction in the tensile stresses at all points along the member. Only when the strain has increased sufficiently to develop the tensile stress equal to the tensile strength of the next weakest point on the basis of the reduced stiffness will the next crack form. This results in the 'saw-toothed' load–deformation and load–crack width relation shown in Fig. 8.33.

8.3.3. Derivation of crack prediction formulae
The development of formulae for the prediction of crack widths given in *clause 4.4.2.4* follows from the description of the cracking phenomenon given above. If it is assumed that all the extension occurring when a crack forms is accommodated in that crack, then, when all the cracks have formed, the crack width will be given by

Clause 4.4.2.4

$$w = S_{rm}\epsilon_m$$

where w is the crack width, S_{rm} is the average crack spacing and ϵ_m is the average strain.

This is simply a statement of compatibility. Since no crack can form within S_0 of an existing crack, this defines the minimum spacing of the cracks. The maximum spacing is $2S_0$ since, if a spacing existed wider than this, a further crack could form. It follows that the average crack spacing will lie between S_0 and $2S_0$. This is frequently assumed to be $1 \cdot 5 S_0$.

It is in the calculation of S_{rm} that the most significant differences arise between the formulae in national codes. The distance S_0, and hence S_{rm}, depends on the rate at which stress can be transferred from the reinforcement, which is carrying all the force at a crack, to the concrete. This transfer is effected by bond stresses on the bar surface. If it is assumed that the bond stress is constant along the length S_0, and that the stress will just reach the tensile strength of the concrete at a distance S_0 from a crack, then

$$\tau \pi \phi S_0 = A_c f_{ct}$$

where τ is the bond stress, A_c is the area of concrete and f_{ct} is the tensile strength of concrete.

Taking $\rho = \pi \phi^2 / 4 A_c$ and substituting for A_c gives

$$S_0 = f_{ct} \pi \phi / 4 \rho$$

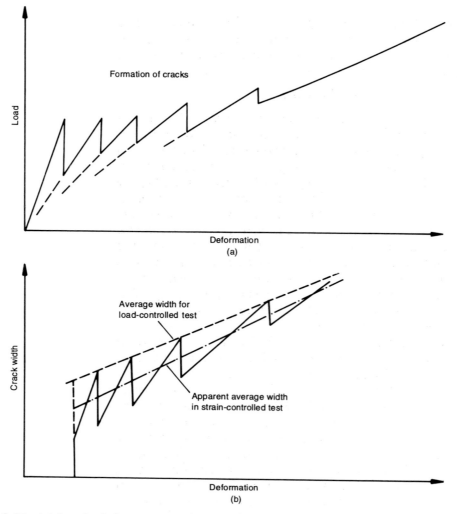

Fig. 8.33. (a) Load–deformation response of a member subjected to steadily increasing deformation; (b) crack width–deformation response for a member subjected to steadily increasing deformation

and from this

$$S_{rm} = 0 \cdot 25k\phi/\rho$$

where k is a constant depending on the bond characteristics of the reinforcement. This is the oldest form of relation for the prediction of crack spacings. More recent studies have shown that the cover also has a significant influence, and that a better agreement with test results is obtained from an equation of the form

$$S_{rm} = kc + 0 \cdot 25k_1\phi/\rho$$

where c is the cover.

This formula has been derived for members subject to pure tension. In order to be able to apply it to bending, it is necessary to introduce a further coefficient k_2, and to define an effective reinforcement ratio ρ_r. These take account of the different form of the stress distribution within the tension zone and the fact that only part of the section is in tension. These factors can be derived empirically from tests. In deriving the formula in EC2, it was felt that to include the cover explicitly in the formula would encourage designers to reduce the cover to a minimum, since this reduced the crack width. To avoid this, the term kc was replaced with a constant figure of 50 mm. This was shown to make only a slight difference to the accuracy of the prediction. The resulting formula is

$$S_{rm} = 50 + 0 \cdot 25 k_1 k_2 \phi / \rho_r$$

where k_1 is a coefficient taking account of the bond properties of the reinforcement (a value of $0 \cdot 8$ is taken for high-bond bars and one of $1 \cdot 6$ for smooth bars) and k_2 is a coefficient depending on the form of the stress distribution. A value of $0 \cdot 5$ is taken for bending and one of 1 for pure tension. Intermediate values can be obtained from the relation $k_2 = (\epsilon_1 + \epsilon_2)/2\epsilon_1$ where ϵ_1 and ϵ_2 are the greater and lesser tensile strains respectively at the faces of the member. ρ_r is the effective reinforcement ratio where A_s is the area of tension reinforcement contained within the effective area of concrete in tension $A_{c.eff}$. This is the area of concrete in tension surrounding the reinforcement of a depth equal to $2 \cdot 5$ times the distance from the tension face of the member to the centroid of the tension reinforcement. A figure in EC2 gives definitions for other, less typical cases.

The other parameter in the crack-width equation is the average strain ϵ_{sm}. This is obtained from *equation (4.81)*. Its derivation is considered in section 8.4 below, where the calculation of deflections is considered.

The equation for crack width gives an estimate of the mean crack width. In design, it is not the mean width that is required, but a width with a considerably lower likelihood of occurrence. EC2 aims to predict a 'characteristic' width (a width with a 5% probability of being exceeded). Studies of the frequency distributions of crack widths show that a reasonable estimate of the characteristic width can be obtained by multiplying the mean width by $1 \cdot 7$. There remains a situation where the formula described above can lead to a significant overestimate of the likely cracking. The reason for this may be understood by considering the element shown in Fig. 8.34. This is an unreinforced element subjected to an axial load applied at an eccentricity sufficiently large for part of the section to be in tension. If the load is sufficient, the section will crack. The formation of this crack will not result in failure, but merely in a redistribution of stresses locally to the crack. Some distance away from this crack, the stresses will remain unaffected by the crack. It is found that the crack affects the stress distribution within a distance roughly equal to the height of the crack on either side of the crack. Thus, by the same arguments used above, the spacing of the cracks can eventually be expected to be between the crack height and twice the crack height. This leads to the relation

$$S_{rm} = h - x$$

where h is the overall depth of the section and x is the depth of the neutral axis.

This formula not only applies to members subject to axial compression, but also to any situation where the cracks, when they form, do not pass right through the section. The effect of bonded reinforcement in the section is almost always to give a calculated crack spacing much smaller than that given by the above equation. The equation does, however, give a maximum, or limiting, value for the spacing, and there are a number of practical situations where it can be used to advantage. One particular case is that of a prestressed beam without any bonded normal reinforcement. The bond of prestressing tendons or wires is often much inferior to that on normal high-bond reinforcement. A safe estimate of the crack spacing, and hence of the crack width, can be made by treating the prestress as an external load, calculating the depth of the tension zone under the loading considered, and applying the above formula. *Clause 4.4.2.4(8)* permits this procedure.

Clause 4.4.2.4(8)

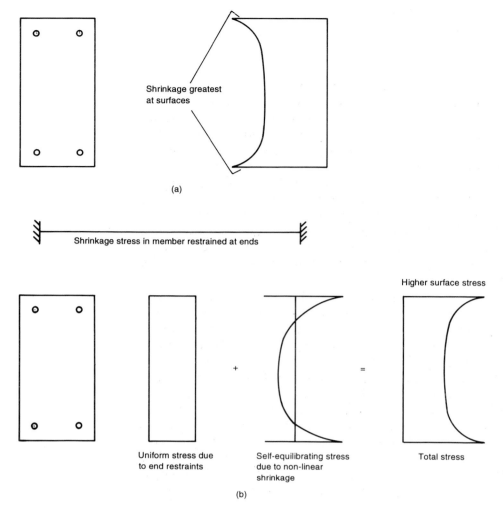

Fig. 8.34. Development in a reinforced concrete member of (a) shrinkage, (b) shrinkage-induced stresses

8.3.4. Minimum areas of reinforcement

8.3.4.1. GENERAL PRINCIPLES. An essential condition for the validity of the formulae for crack-width calculation is that the reinforcement remains elastic. If the reinforcement yields, the deformation becomes concentrated at the crack where the yielding is occurring, which is not the condition considered in the derivation of the formulae. The condition immediately after formation of the first crack is particularly critical. Should formation of the first crack lead to yield, then only a single crack will form and all the deformation will be concentrated at this crack. The principles involved here can again be most easily demonstrated by considering a member subjected to pure tension.

The force necessary to cause the member to crack is given by

$$N_r = A_c f_{ct}$$

where N_r is the cracking load, A_c is the area of concrete and f_{ct} is the tensile strength of the concrete.

The strength of the steel is $A_s f_y$. Thus, for the steel not to yield on first cracking and hence for spread cracking to develop

SERVICEABILITY LIMIT STATES

$$A_s f_y > A_c f_{ct}$$

or

$$A_s > A_c f_{ct}/f_y$$

This provides the minimum reinforcement area required for controlled cracking. It can be shown that where the cracking is caused exclusively by loading this limitation is unimportant, since no cracks will form under service conditions. However, if tensile stresses may be generated by restraint of imposed deformations such as shrinkage or thermal movements, it is essential to ensure that at least the minimum reinforcement area is provided. This result is true for flexure as well as for pure tension. Since it is only in rare cases that it can be said with confidence that there is no possibility of stresses arising from restraint, it is advisable to apply the rules for minimum steel area in the great majority of practical cases.

While the principles illustrated above for pure tension apply generally, the actual equations will differ for different types of member. To avoid excessive complexity in the rules, a factor k_c is introduced into the above relation to adjust for different forms of stress distribution.

A further factor k is included to allow for the influence of internal self-equilibrating stresses. These arise in cases where deformations of the concrete in the member itself are restrained. The most common forms of such restrained deformation result from either shrinkage or temperature change. These effects do not occur uniformly throughout the section, but will occur more rapidly near the member surface. As a result, the deformation of the surface concrete will be restrained by the interior concrete, and higher tensions will be developed near the surface. This is illustrated in Fig. 8.34. The higher tensile stresses at the surface will lead to the occurrence of cracking at a lower load than would be predicted on the basis of a linear distribution of stress across the section. Since the cracking load will be smaller than predicted, a smaller amount of reinforcement will be necessary to ensure that controlled cracking occurs. The function of the factor k is thus to reduce the minimum steel area in cases where such non-linear stress distributions can occur.

Finally, it is necessary to consider what value should be chosen for the tensile strength of the concrete f_{ct}. It will be seen from the equations above that the minimum area of reinforcement is proportional to the tensile strength of the concrete. It thus seems logical to take as the tensile strength of the concrete an estimate of its likely maximum value. Such a value is not easy to establish and, furthermore, would lead to levels of reinforcement that would be considered impractically high. For this reason, a more pragmatic approach has been adopted and the provisions of EC2 aim to give a minimum area that is not too different from current practice.

The above considerations lead to the formula given in *clause 4.4.2.2* for minimum reinforcement

Clause 4.4.2.2

$$A_s \geq k_c k f_{ct.eff} A_c / \sigma_s$$

where k_c is a coefficient taking account of the form of the loading, k is a coefficient taking account of the possible presence of non-linear stress distributions, and $f_{ct.eff}$ is the tensile strength of the concrete effective at the time when the cracks first form. Except where the cracks can be guaranteed to form at a very early age, it is suggested that the value of $f_{ct.eff}$ chosen should not be less than 3 N/mm². A_c is the area of concrete in tension immediately before the formation of the first crack, and σ_s is the steel stress, which can generally be taken as the yield strength of the steel although, as seen below, there are cases where it may be more convenient to use other, lower values.

8.3.4.2. INTERPRETATION OF THE FORMULA.
EC2 provides a considerable amount of guidance on how to arrive at values for the coefficients k and k_c and the other parameters in the equation. Nevertheless, there is room for doubt on what values should be chosen in particular cases. Some of these are now discussed.

8.3.4.2.1. Values for k_c. k_c takes values of $1 \cdot 0$ for pure tension, $0 \cdot 4$ for pure flexure and between 0 and $0 \cdot 4$ for prestressed beams. EC2 states in *clause 4.4.2.2(7)* that k_c may be taken as zero if

Clause 4.4.2.2.(7)

(a) under a rare combination of loads, the whole section remains in compression, or
(b) under the action of the estimated value of prestress, the depth of the tension zone calculated on the basis of a cracked section under the loading leading to the formation of the first crack does not exceed the lesser of $h/2$ and 500 mm.

Values of k_c may be interpolated between 0 where one of conditions (a) or (b) applies, and $0 \cdot 4$ for pure flexure with no prestress. How this interpolation might be carried out is not explained. The question of how to interpolate between pure flexure and pure tension is also left unexplained. A possible logical way of solving this problem is developed below.

It is shown above how the formula for minimum reinforcement could be derived for pure tension. Similar but more complex calculations can be carried out for bending, or bending combined with compression or tension. For rectangular sections, the results of these calculations can conveniently be expressed in graphical form (Fig. 8.35). This is an alternative presentation of the formula for minimum reinforcement, which does not include k_c directly. It can, however, be shown to give the same answer for the two specific cases for which k_c is given in EC2. For pure bending, $NK/bhf_{ct.eff} = 0$ and Fig. 8.35 gives

$$K\sigma_s/f_{ct.eff} = 0 \cdot 2$$

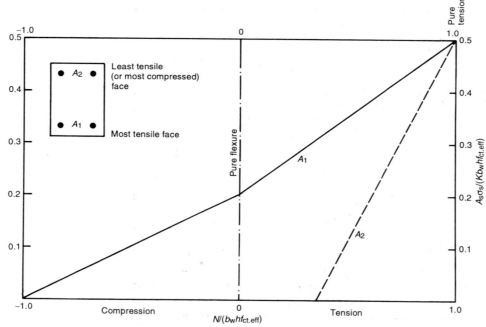

Fig. 8.35. Design chart for assessing minimum areas of reinforcement

For a rectangular section A_{ct}, the area of concrete in tension just before the formation of the first crack is equal to $bh/2$. Hence, $bh/A_{ct} = 2$. Substituting for this, introducing $k_c = 0.4$ and rearranging will show that the equation and the chart give identical results in this case. Similarly, for pure tension, the chart requires an area of reinforcement of $0.5bhkf_{ct.eff}/\sigma_s$ in the top and bottom of the

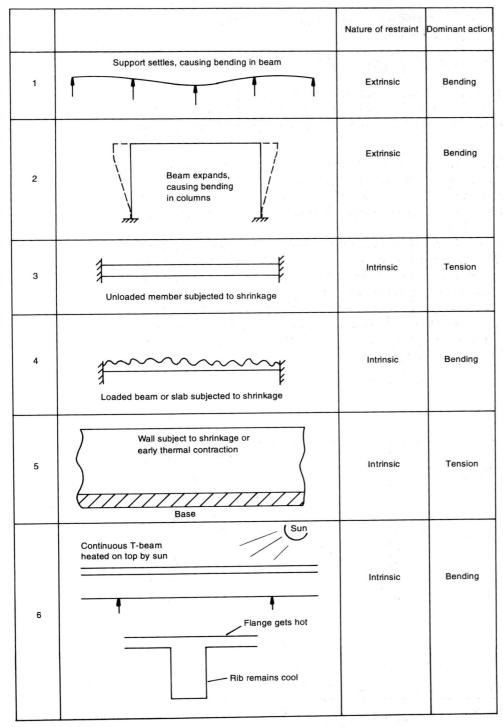

Fig. 8.36. *Intrinsic and extrinsic restrained imposed deformations*

section. The formula requires a total of $1 \cdot 0bhkf_{ct.eff}/\sigma_s$, which is the same. Fig. 8.35, however, has the advantage of providing appropriate reinforcement areas for combined bending and tension and combined bending and compression. Prestress can be treated as an externally applied compression, and hence the chart will conveniently deal with the areas left ambiguous in EC2.

8.3.4.2.2. Values for k. The basic problem here is to establish whether the deformations are 'extrinsic' or 'intrinsic'. Extrinsic deformations are imposed from outside the element in question. An example of an extrinsic deformation is the settlement of a foundation. In cases such as this, the question of non-linear distributions of strain across a section does not arise, and k must take a value of $1 \cdot 0$. Intrinsic deformations are generated within the member considered. Deformations caused by shrinkage of the member or a change in temperature of the member fall into this category. Fig. 8.36 gives examples of common causes of extrinsic and intrinsic deformations, to assist in the choice of a suitable value of k.

8.3.5. Checking cracking without direct calculation

Flexural or tension cracking can normally be controlled by application of simplified detailing rules. These take the form of tables of maximum bar diameters of maximum bar spacings. The tables *(Tables 4.11 and 4.12)* were derived from parameter studies carried out using the crack width calculation formulae described above. It should be noted that, for load-induced cracking, *either Table 4.11 or*

Clause 4.4.2.3

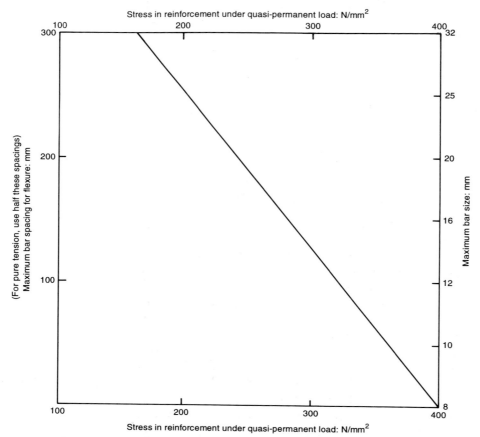

Fig. 8.37. Maximum bar sizes and spacings for crack control

Table 4.12 may be used: it is *not* necessary to satisfy both tables. For cracking due to restrained imposed deformations, *Table 4.11* should be used. For reinforced concrete, the tables can be presented graphically as in Fig. 8.37. For load-induced cracking, the steel stress is the stress under the quasi-permanent combination of loading. This may be calculated approximately from

$$f_{sqp} = f_{yk}N_{qp}/1 \cdot 15N_{ud}$$

where f_{sqp} is the steel stress under the quasi-permanent load, N_{qp} is the quasi-permanent load and N_{ud} is the design ultimate load.

For prestressed concrete beams containing bonded untensioned reinforcement, it is possible to make an estimate of the stress in the untensioned steel from Fig. 8.35. This is done by calculating a ficticious value for $f_{ct.eff}$ equal to the tensile stress in the concrete at the beam soffit, on the assumption that the concrete is uncracked. A value of $N/bhf_{ct.eff}$ and a value for $\rho\sigma_s/f_{ct.eff}$ can now be calculated. Substituting for ρ and $f_{ct.eff}$ will give a value for σ_s. Example 8.2 illustrates the procedure.

Example 8.2
Carry out a check for cracking for a prestressed T-beam having a rib breadth of 250 mm, an overall depth of 600 mm, a flange breadth of 1000 mm and a flange depth of 180 mm. The prestress N after allowance for losses is 700 kN at an eccentricity of 280 mm, and the moment under the frequent combination of loads is 330 kN m.

It can be found from standard tables that the section modulus for stresses at the bottom of the section is $0 \cdot 222bh^2 = 20 \times 10^6$ and the area of the section is 285×10^3 mm². The stress at the soffit of the section is thus given by

$$f_{ct.eff} = 700/285 + 700 \times 0 \cdot 28/20 - 330/20 = -4 \cdot 09$$

Hence
$$N/Af_{ct.eff} = 700/285/4 \cdot 09 = 0 \cdot 6$$

From Fig. 8.10
$$A_s\sigma_s/b_rhf_{ct.eff} = 0 \cdot 08$$

Hence
$$A_s\sigma_s = 0 \cdot 08 \times 250 \times 600 \times 4 \cdot 09 = 49\,080$$

Assume that 12 mm bars are chosen. This gives $A_s = 226$ mm² and hence $f_s = 217$ N/mm². For this stress, *Table 4.11* gives a maximum bar diameter of between 12 and 16 mm, so the design is satisfactory. *Table 4.12* gives a maximum bar spacing of about 130 mm, which would be difficult to achieve with only two bars. However, since *Table 4.11* has been satisfied, this does not matter.

It is, of course, possible to calculate the stress in the reinforcement rigorously on the basis of a cracked section. This would have given a significantly lower stress if proper allowance had been made for the prestressing tendons in the section, so the proposed method is on the safe side. Use of the rigorously calculated stress and the formulae given for the calculation of crack width results in a width below 0·2 mm. *Clause 4.4.2.3(2)* states that the increase in stress in the tendons *should* be ignored when calculating the steel stress for use in the tables. It is suggested that this provision was included to simplify the calculation, and that it should more correctly read '*may* be ignored'. It *should*, however,

Clause 4.4.2.3(2)

be ignored when the procedure outlined above is used, because otherwise an unconservative result may be obtained.

EC2 gives no way of using the tables where there is no untensioned reinforcement in the section. From an interpretation of the rules given in *clause 4.4.2.4(4)* it seems a reasonable approximation to calculate a steel stress as described above. This will be a change in stress in the tendons beyond the condition where the concrete at the level of the tendons is at zero stress. To allow for the inferior bond of the tendons, the bar diameters given by *Table 4.11* or the bar spacings from *Table 4.12* should be divided by 2·5.

Clause 4.4.2.4(4)

Table 4.13 provides rules for the spacing of stirrups for the control of the widths of shear cracks. In general it should be found that crack control is not a serious limiting factor for defining stirrup spacings, but *Table 4.13* may occasionally have an influence. *Table 4.13* has been arrived at from the following approximate considerations.

Equation (4.83) gives the following expression for the crack spacing where the cracks form at a significant angle to the direction of the reinforcement

$$S_{rm} = 1/(\cos\theta/S_{rmx} + \sin\theta/S_{rmy})$$

If it is assumed that shear cracks form at 45° to the axis of the beam, then $\cos\theta = \sin\theta = 0.707$. As a conservative approximation, the crack spacing at mid-height of the section can be assumed to be $(h-x)$ (applying *clause 4.4.2.4(8)*). The crack spacing perpendicular to the stirrups can be assessed by the normal formula. This gives a means of calculating the spacings of shear cracks which can be used as the basis for a parameter study.

Clause 4.4.2.4(8)

The stress in the stirrups, and hence the strain, may be assessed roughly from

$$f_{sv} = (V_{ser} - V_{cds})/\rho_w b_w z$$

where f_{sv} is the stress in the stirrups, V_{ser} is the shear force under service load and V_{cds} is the shear capacity of concrete under service conditions.

It seems reasonable to assume that V_{ser} is about V_{Sd} for the ultimate limit state divided by 1·4, and that V_{cds} can be calculated roughly from the relation given for V_{Rd1} but with the safety factor of 1·5 removed and based on the mean tensile strength of the concrete rather than the lower characteristic. Introducing these assumptions and resolving the stress to a value perpendicular to the cracks gives

$$f_{sv} = (V_{Sd} - 3V_{Rd1})/1.26\rho_w b_w d$$

The vertical strain can be obtained directly from this relation.

The longitudinal strain may reasonably be taken as the bending strain at a height midway between the main reinforcement and the neutral axis. The worst condition for shear cracks will be near the supports of a continuous beam, where the stress in the tension reinforcement under service loads may be assumed to be in the region of $0.7f_{yd}$. The strain at the height considered can thus be estimated to be

$$x = 0.7f_{yd}/2 \times 200\,000$$

An extensive parameter study was carried out using these assumptions, the values in *Table 4.13* give a lower bound to the results of this study.

It should be noted that there are further limitations on stirrup spacing in *clause 5.4.2.2*.

Clause 5.4.2.2

8.3.6. Checking cracking by direct calculation

The formula for calculating the design crack width is discussed in section 8.3.3 above.

8.4. Control of deflections

8.4.1. General

Deflection control is dealt with in *clause 4.4.3*, which deals with the control of deflection using simple 'deemed to satisfy' rules, and in *Appendix 4*, which deals with the calculation of deflections. It was the view of the drafting committee that, in general, the calculation of deflections gave an unwarranted impression of precision in what was a very uncertain process. It was felt that the use of simple rules, such as limits to span/effective depth ratios, was a perfectly adequate approach for all normal situations. Hence the relegation of the calculation of deflections to an appendix. In this chapter, however, the intentions of EC2 and the derivation of the simplified methods will be understood better if the calculation of deflections is covered first. In design for all serviceability limit states, there are four necessary elements:

Clause 4.4.3
Appendix 4

(a) criteria defining the limit to satisfactory behaviour
(b) appropriate design loads
(c) appropriate design material properties
(d) means of predicting behaviour.

Each of these is now considered in turn, with reference to deflections.

8.4.2. Deflection limits

The selection of limits to deflection which will ensure that the structure will be able to fulfil its required function is a complex process, and it is not possible for a code to specify simple limits that will meet all requirements and still be economical. For this reason, EC2 makes it clear (in *clause 4.4.3.1*) that it is the responsibility of the designer to agree suitable values with the client, taking into account the particular requirements of the structure. Limits are suggested in EC2 but these are for general guidance only; it remains the responsibility of the designer to check whether these are appropriate for the particular case considered or whether some other limits should be used.

Clause 4.4.3.1

Two basic issues are considered to influence the choice of limits, i.e. appearance and function.

Appearance is important because it is found that the occupants of structures find it upsetting if the floors appear to be sagging. Some research has been carried out on this highly subjective aspect of deflection control, and it is generally accepted that sag will be unnoticeable provided the central deflection of a beam relative to its supports is less than about span/250.

Function is more difficult to cover, as the range of possibilities is large. Examples of situations where deflections may lead to impairment of function are

(a) deflection of beams or slabs leading to cracking of partitions supported by the member considered
(b) deflections causing doors to jam or windows to break
(c) varying deflections leading to misalignment of apparatus or machinery mounted on the member considered
(d) damage to brittle finishes
(e) unacceptable vibrations or an upsetting feeling of 'liveliness' in the structure.

Of the above list, damage to brittle partitions is probably the most common problem and the one that code limits are generally formulated to avoid. A number of typical cases of damage to partitions caused by deflections are illustrated in Fig. 8.38. The width of the resulting cracks is frequently large, 2–5 mm not being uncommon. It is difficult to define limits that will avoid such cracking entirely,

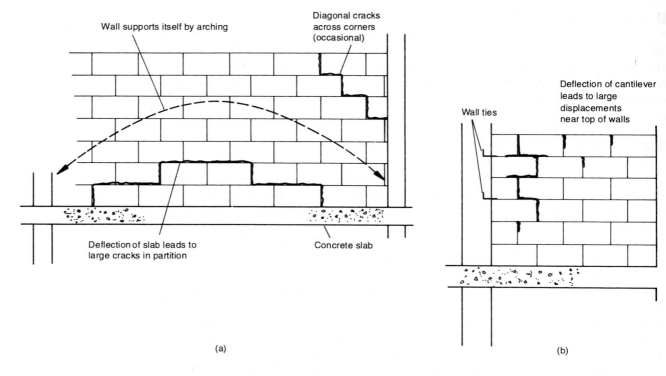

Fig. 8.38. Damage to partitions due to deflection of (a) beam, (b) cantilever

and surveys have found cracking with deflections below span/1000. Such small limits to deflection are generally considered uneconomic, and many codes (including EC2) suggest a higher limit of span/500. Some codes suggest limits as high as span/300. The limit chosen in a particular case may depend on the explanation to the client of the consequences of various options. For example, a deflection limit of span/300 may be quite acceptable if the client is prepared to accept that significant cracking may occur in partitions after a year or so but that, if the cracks are filled up, they will probably not reappear. This may be a considerably more economical approach than increasing the depth of the members to a level that will ensure no cracking. The limit can, in any case, be relaxed or ignored if the partitions are either flexible enough to accommodate the deflection or detailed so that some relative movement between the partition and the supporting member can occur.

8.4.3. Design loads

When the overall deflection is being checked to ensure that the appearance is not impaired, EC2 takes the view that, since in concrete structures the dominant part of the deflection is that occurring under the permanent loads, it is necessary only to check the deflection under the quasi-permanent combination of loads. The possibility that the deflection may occasionally exceed the calculated value is considered unimportant in this case.

EC2 is not specific about the loading that should be used when checking for damage to partitions but, again, since the major part of the deflection will be due to creep and shrinkages, and the calculation of deflections is a highly approximate process, it seems reasonable to use the quasi-permanent load. It should be pointed out, however, that *Appendix 4.2(5)* suggests the use of the frequent combination of loads, assuming for simplicity that this load is of long duration.

Appendix 4.2(5)

8.4.4. Material properties

Clause 4.2P(1) states that the calculation method should represent the true behaviour of the structure to an accuracy appropriate to the objectives of the calculation. This suggests that the material properties assumed in the calculation should reflect a best estimate of their values rather than a lower bound. This is confirmed by the provisions of *section 4.3*, where the average values of tensile strength and modulus of elasticity are used for concrete (f_{ctm} and E_{cm}).

Clause 4.2P(1)

section 4.3

8.4.5. Model of behaviour

8.4.5.1. SHORT-TERM BEHAVIOUR.
Figure 8.39 gives an idealized picture of the load–deformation characteristics of a reinforced concrete beam. It is convenient to consider the curve to be made up of three phases.

Phase 1 (uncracked): in this phase, the tensile strength of the concrete has not been exceeded. The section behaves elastically and its behaviour can be predicted on the basis of an uncracked section, but with allowance made for the reinforcement.

Phase 2 (cracked): in this phase, the concrete has cracked in tension. The concrete in compression and the reinforcement may, however, be considered to remain elastic. The behaviour of the tension zone is complex. At a crack the concrete in tension carries no stress. However, between cracks, bond transfers stress from the reinforcement to the concrete so that, with increasing distance from a crack, the tension carried by the concrete increases. The behaviour illustrated in Fig. 8.39 for this phase reflects the average state, where the tension zone carries some tension. This is perfectly satisfactory for deflection calculations, since the calculation of the deflection requires the integration of the behaviour over the length of the beam.

Phase 3 (inelastic): in this phase, the steel has yielded or the concrete is stressed to a level where the assumption of elasticity ceases to be reasonable, or both. This phase is generally reached only at loads well above those likely to occur in normal service, and so is not of interest for serviceability calculations.

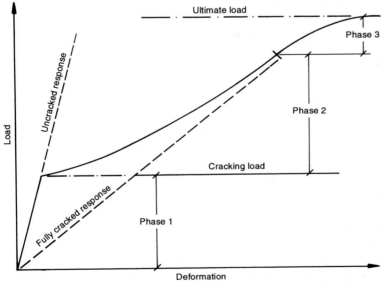

Fig. 8.39. *Idealized load–deformation characteristics of a reinforced concrete member*

It is the behaviour in phase 2 that causes difficulties, and all current design approaches are empirical and approximate.

It is possible to go some way towards defining the characteristics of a prediction method, since certain limits to behaviour can be defined, as follows.

(a) At the instant of cracking, when the tensile strength of the concrete is just attained, the response of the member must lie on the phase 1 line.

(b) Since cracking effectively reduced the stiffness of the member, behaviour after cracking must lead to curvatures greater than the phase 1 curvature.

(c) The maximum possible curvature corresponds to the condition where the concrete in tension carries absolutely no stress. This is the response that would be calculated on the basis of a cracked transformed section. The curvature corresponding to this is indicated in Fig. 8.39.

In practice, experiments show that, at the cracking moment, the behaviour lies on the phase 1 line and, as the load is increased above the cracking load, the response tends towards the fully cracked response. This would be expected, since increase in cracking and bond slip in the region of the cracks leads to an increasing loss of effectiveness of the concrete in tension. Many formulae have been developed that satisfy the basic requirements set out above, and there is probably not much difference between them. The method given in *Appendix 4* is one such, and has the practical advantage of being relatively easy to apply. *Appendix 4*

The basic concept of the method is illustrated in Fig. 8.40. Considering a length of a beam bounded by two cracks, it is assumed that some length close to the cracks is fully cracked while the remainder is uncracked. Considering the whole crack spacing s, it will be seen that the length ξs is considered fully cracked and the length $(1-\xi)s$ is considered uncracked.

For the simple case of pure flexure, the rotation over the length s is given by

$$\theta = \xi s(1/r_2) + (1-\xi)s(1/r_1)$$

Hence, the average curvature is given by

$$1/r_m = \theta/s = \xi(1/r_2) + (1-\xi)(1/r_1)$$

The same principle can be adopted for the calculation of any parameter relating to behaviour under any condition of loading. Thus, for example, the average steel strain may be calculated using

$$\epsilon_{sm} = (1-\xi)\epsilon_{s1} + \epsilon_{s2}\xi$$

In the above equations, the subscripts 1 and 2 indicate behaviour calculated assuming the section to be uncracked and fully cracked respectively.

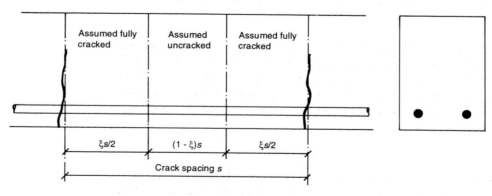

Fig. 8.40. *Model of behaviour in phase 2*

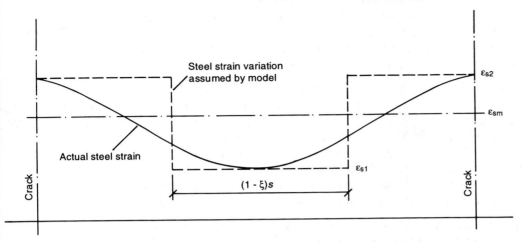

Fig. 8.41. *Alternative visualization of model of behaviour in phase 2*

An alternative way of visualizing the idealization is shown in Fig. 8.41 for the strain in the reinforcement.

So far, the model is simply a convenient way of expressing the condition that the actual behaviour after cracking must lie between that calculated for an uncracked section and that calculated for a fully cracked section. Any actual result could be simulated by the suitable choice of the distribution coefficient ξ. Empiricism now enters the process in order to define a suitable expression for the distribution coefficient ξ. Any suitable expression must have the property that, at the cracking load, it must take the value 0 and, with increasing load, the value must approach 1·0. EC2 adopts the expression

$$\xi = 1 - \beta_1(\sigma_{sr}/\sigma_s)^2$$

where β_1 is a coefficient characterizing the bond properties of the reinforcement (it is proposed that for plain bars $\beta_1 = 0 \cdot 5$, and for ribbed bars $\beta_1 = 1 \cdot 0$), σ_{sr} is the steel stress calculated on the basis of a cracked section under the loading that just causes the tensile strength of the concrete to be attained at the section considered, and σ_s is the steel stress under the loading considered, calculated on the basis of a cracked section.

8.4.5.2. LONG-TERM BEHAVIOUR.
There are three factors to be considered in calculating the long-term behaviour: creep, shrinkage and, in cracked sections, reduction of the tensile stresses in the concrete in the tension zone due to the spreading with time of cracking and local bond failure. These are now considered in turn.

8.4.5.2.1. Creep.
Creep can conveniently be dealt with using the effective modulus method. This is not generally considered to be the most accurate method of modelling creep, but it has the advantage of simplicity. In view of the approximate nature of deflection calculations, a more accurate but less convenient method would not be justified. The effective modulus of elasticity of the concrete, taking account of creep, is given by

$$E_{c.eff} = E_{cm}/(1 + \phi)$$

where $E_{c.eff}$ is the effective modulus of elasticity of the concrete, E_{cm} is the short-term modulus of elasticity of the concrete, and ϕ is the creep coefficient. Values for the creep coefficient are discussed in section 8.1.2.3 above.

8.4.5.2.2. Shrinkage. Uniform shrinkage of an unrestrained, unreinforced member will simply lead to an overall shortening, without either curvature or stresses being induced. Reinforcement, which does not shrink, will restrain the shrinkage to some degree. This will lead to compression in the reinforcement and tension in the concrete. Where the reinforcement within the section is unsymmetrical, the restraint provided by the reinforcement will also be unsymmetrical and a curvature will be induced. In shallow members this curvature can be large enough to produce significant deflections that will need to be taken into account in any calculations.

The method given in EC2 for the calculation of shrinkage deformations can be derived for uncracked sections as follows.

Consider, for simplicity, the simply supported rectangular beam shown in Fig. 8.42. If this beam is constrained to shorten by the amount of the free shrinkage, a compressive stress $\epsilon_{cs}E_s$ will be induced in the reinforcement. This is equivalent to a force N_{cs} equal to $\epsilon_{cs}E_sA_s$, where ϵ_{cs} is free shrinkage strain, E_s is the modulus of elasticity of the reinforcement and A_s is the area of the reinforcement.

If the system is now released, the beam will deform under the released force in the steel. This leads to a curvature given by

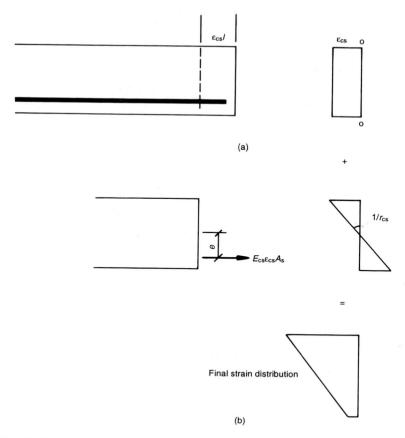

Fig. 8.42. *Deformation due to shrinkage of a rectangular beam: (a) if free, concrete would shorten by $\epsilon_{cs}l$; (b) shortening of concrete would induce a force in the reinforcement and hence develop an internal moment, hence a curvature*

$$1/r_{cs} = N_{cs}e/E_cI_1$$
$$= \epsilon_{cs}E_sA_se/E_cI_1$$
$$= \alpha_e\epsilon_{cs}S/I_1$$

where α_e is the modular ratio ($= E_s/E_c$), e is the eccentricity of reinforcement, I_1 is the second moment of area of the uncracked section, and S is the first moment of area of the reinforcement about the centroid of the concrete section ($= A_se$ for a singly reinforced section).

In the above equations, E_c should clearly be an effective value allowing for the effects of creep, since shrinkage is a long-term effect.

The extension of this method to cover cracked sections is speculative, but seems reasonable. It is therefore suggested simply that, for cracked sections, I_2, the second moment of area of the cracked section, be substituted for I_1 and that S be also calculated for a cracked section in order to calculate the appropriate value of shrinkage curvature for the fully cracked state. The actual curvature is then calculated using the distribution coefficient approach set out in section 8.4.5.1 above.

Values for free shrinkage are considered in section 8.1.2.4 above.

8.4.5.2.3. Bond deterioration.

The effect of sustained loading or repeated loading is to reduce the effectiveness of the transfer of stress from the reinforcement to the concrete between cracks. This effect can be seen to be modelled by increasing the distribution coefficient ξ. This is conveniently done by introducing a second coefficient β_2 into the equation for the distribution coefficient, which then becomes

$$\xi = 1 - \beta_1\beta_2(\sigma_{sr}/\sigma_s)^2$$

β_2 is a coefficient taking account of duration of loading or of repeated loading. It takes values of 1 for monotonic short-term loading and 0·5 for long-term or repeated loads.

Example 8.3

The long-term curvature is to be calculated for the mid-span section of a beam whose cross-sectional details are shown in Fig. 8.43. The moment under the combination of loads considered is 65 kN m. It may be assumed that the tensile strength of the concrete f_{ctm} is 2·2 N/mm², the elastic modulus E_c is 29 kN/mm², the creep coefficient is 2·0 and the free shrinkage strain is 300 × 10⁻⁶.

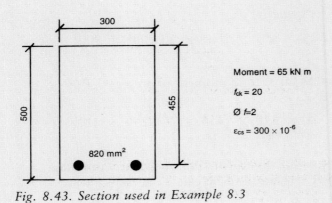

Fig. 8.43. Section used in Example 8.3

The cracking moment is given by
$$M_{cr} = f_{ctm}bh^2/6 = 2\cdot2 \times 300 \times 500^2/6 \times 10^{-6} \text{ kN m}$$
$$= 27\cdot5 \text{ kN m}$$

Since this is much less than the applied moment, the section is clearly cracked.
The effective modulus of elasticity of the section is given by
$$E_{c.eff} = E_c/(1 + \phi) = 29/(1 + 2) = 9\cdot67 \text{ kN/mm}^2$$
The curvature of the uncracked section is given sufficiently accurately as
$$1/r_1 = M/EI = 65 \times 10^6 \times 12/(9\cdot67 \times 10^3 \times 300 \times 500^3)$$
$$= 2\cdot15 \times 10^{-6}$$

In the above calculation, no allowance has been made for the influence of the reinforcement on the second moment of area of the uncracked section. This could be taken into account, but the difference in the final result would not be significant.

The effective modular ratio $\sigma_e = E_s/E_{c.eff} = 20\cdot68$. This leads to a transformed reinforcement ratio of
$$\sigma_e A_s/bd = 20\cdot68 \times 820/300 \times 455 = 0\cdot124$$

Fig. 8.5 gives $x/d = 0\cdot39$ and hence $x = 177$ mm.
The steel stress can now be calculated as
$$M/A_s(d-x/3) = 65 \times 10^6/[820(455 - 177/3)] = 200 \text{ N/mm}^2$$
From this, the curvature of the fully cracked section can be calculated as
$$1/r_2 = s/(d-x) = 200/200\,000\,(455 - 177) = 3\cdot6 \times 10^{-6}$$
It is now necessary to calculate the distribution factor ξ.

The steel stress at the cracking moment can be obtained *pro rata* from the stress under the design moment as
$$200 \times M_{cr}/M = 200 \times 27\cdot5/65 = 84\cdot6 \text{ N/mm}^2$$
For deformed bars and long-term loads, $\beta_1\beta_2 = 0\cdot5$, hence
$$\xi = 1 - 0\cdot5\,(84\cdot6/200)^2 = 0\cdot91$$
$$1/r = \xi(1/r_2) + (1 - \xi)(1/r_1)$$
$$= (0\cdot91 \times 3\cdot6 + 0\cdot09 \times 2\cdot15) \times 10^{-6} = 3\cdot47 \times 10^{-6}$$

The shrinkage curvature for a cracked section and an uncracked section is now calculated.

For the uncracked section, this is given by
$$1/r_{cs1} = 300 \times 10^{-6} \times 20\cdot68 \times (820 \times 205) \times 12/300 \times 500^3$$
$$= 0\cdot33 \times 10^{-6}$$

For the cracked section, Fig. 8.5 gives the second moment of area as $0\cdot066 \times 300 \times 500^3 = 250 \times 10^8$. Hence
$$1/r_{cs2} = 300 \times 10^{-6} \times 20\cdot68 \times 820 \times (455-177)/250 \times 10^8$$
$$= 0\cdot57 \times 10^{-6}$$

Use of the same distribution factor as for the calculation of the curvature due to load enables the actual shrinkage curvature to be calculated as
$$1/r_{cs} = (0\cdot91 \times 0\cdot57 + 0\cdot09 \times 0\cdot33) \times 10^{-6} = 0\cdot55 \times 10^{-6}$$

The total curvature is thus

$$(3\cdot 47 + 0\cdot 55) \times 10^6 = 4\cdot 02 \times 10^{-6}$$

It will be seen that, even with the relatively low reinforcement ratio used in Example 8.3, the calculated curvature does not differ much from that calculated assuming a fully cracked section. For sections with higher percentages of reinforcement, it is probably sufficiently accurate simply to calculate long-term curvatures on the basis of a fully cracked section.

8.4.5.3. CALCULATION OF DEFLECTIONS FROM CURVATURES.

The deflection of a member is calculated by double integration of the curvature over the length of the beam and the introduction of appropriate boundary conditions.

The most general way of achieving this is by calculating the curvature at intervals along the member and then calculating the deflection by numerical integration. Using this method, it is possible to take account of the complex relation between moment and curvature due to the development of cracking, and also the effects of varying steel percentage or section shape. The procedure may be iterative for indeterminate beams, since the moment field calculated using normal methods of analysis is unlikely to be the correct one when allowance has been made for the varying stiffness from point to point along the beam due to the development of cracking.

A calculation of this type can conveniently be set out in tabular form, and is well suited to the use of a spreadsheet program. Such an approach is illustrated in Example 8.4.

Example 8.4
The deflection of a simply supported beam of span $5\cdot 1$ m supporting a uniformly distributed load of 20 kN/m is to be calculated. The section details, material properties, cracked neutral axis depth and second moment of area are as given in Example 8.3. The moment at any section is given by

$$M = nlx/2 - nx^2/2$$

As in Example 8.3, the mid-span moment can be calculated to be 65 kN m. The curvatures at tenth points along the span are calculated using the method illustrated in Example 8.3 by the spreadsheet given in Table 8.5.

It will be seen that the values in Table 8.5 for mid-span are the same as in Example 8.3. Table 8.6 continues the calculation by carrying out the integration and introducing the boundary conditions. The integrations are carried out using the trapezoidal rule. The first integration, giving the uncorrected rotations at each section, is thus given by

$$\theta_i = \theta_{i-1} + [1/r_i + 1/(r_{i-1})]/2 \times l/10$$

The uncorrected rotations having been established, the uncorrected deflections are obtained by the second integration, given by

$$a_{ui} = a_{u(i-1)} + (\theta_i - \theta_{i-1})/2 \times l/10$$

The required boundary condition is that the deflection should be zero at both supports. This can be introduced by applying a linear transformation to the uncorrected deflections. In this case, there is an uncorrected deflection of $34\cdot 88$ at the right-hand end. The correction is effected by subtracting $34\cdot 88 x/l$ from each uncorrected deflection, where x is the distance from the left-hand support.

This gives the final deflected shape of the beam (given in the extreme right-hand column of Table 8.6).

Table 8.5. Calculation of curvatures at tenth points

Fraction of span	Moment: kN m	$1/r_1 \times 10^6$	$1/r_2 \times 10^6$	Factor	$1/r_{cs} \times 10^6$	$1/r(tot) \times 10^6$
0	0.0	0.00	0.00	0.00	0.33	0.33
0.1	23.4	0.77	1.30	0.31	0.40	1.34
0.2	41.6	1.38	2.31	0.78	0.52	2.62
0.3	54.6	1.81	3.03	0.87	0.54	3.41
0.4	62.4	2.07	3.46	0.90	0.55	3.87
0.5	65.0	2.15	3.60	0.91	0.55	4.02
0.6	62.4	2.07	3.46	0.90	0.55	3.87
0.7	54.6	1.81	3.03	0.87	0.54	3.41
0.8	41.6	1.38	2.31	0.78	0.52	2.62
0.9	23.4	0.77	1.30	0.31	0.40	1.34
1	0.0	0.00	0.00	0.00	0.33	0.33

Table 8.6. Calculation of deflection from curvatures

Fraction of span	$1/r(tot) \times 10^6$	Uncorrected rotation	Uncorrected deflection	Correction	Final deflection
0	0.33	0.0	0.0	0.0	0.0
0.1	1.34	0.4	0.1	3.5	−3.4
0.2	2.62	1.4	0.6	7.0	−6.4
0.3	3.41	3.0	1.7	10.5	−8.8
0.4	3.87	4.8	3.7	13.9	−10.3
0.5	4.02	6.8	6.7	17.4	−10.8
0.6	3.87	8.9	10.7	20.9	−10.2
0.7	3.41	10.7	15.7	24.4	−8.7
0.8	2.62	12.2	21.5	27.9	−6.4
0.9	1.34	13.3	28.0	31.4	−3.4
1	0.00	13.6	34.9	34.9	0.0

8.4.5.4. SIMPLIFIED METHODS OF DEFLECTION CALCULATION. The use of numerical integration is tedious and will be unnecessary in most practical situations. Simpler approaches will normally be adequate.

A simplification that may normally be made is to calculate the curvature at only one point, usually the point of maximum moment, and then assume that the shape of the curvature diagram is the same as the shape of the bending moment diagram. The deflection may then be calculated from

$$a = kl^2(1/r)$$

where k is a constant which depends on the shape of the bending moment diagram. Values for k are given in Table 8.7, taken from the UK Code, BS8110:Part 2:1985.

For a uniformly distributed load on a simply supported beam, k is 0.104. Hence the deflection of the beam used in Example 8.4 above may be calculated as

$$a = 0.104 \times 5.1^2 \times 4.019 = 10.87 \text{ mm}$$

This is very close to the value of 10.77 mm calculated by the more rigorous approach. Errors will be greater where the maximum moment is closer to the cracking moment, but will generally not exceed 10%.

It is suggested that, where the deflection of a continuous beam is being calculated, the critical section at which the curvature is calculated should be the section with

Table 8.7. Values of k for various bending moment diagrams

Loading	Bending moment diagram	k
End moments M and M	Rectangular, M	0.125
Point load W at al from left support, span l	Triangular, $M = Wa(1-a)l$	$\dfrac{3 - 4a^2}{48(1-a)}$ If $a = \dfrac{1}{2}$, $k = \dfrac{1}{12}$
End moment M at one end	Triangular, M	0.0625
Two point loads $W/2$ at al from each support	Trapezoidal, $M = (Wal)/2$	$0.125 - \dfrac{a^2}{6}$
UDL q over full span	Parabolic, $(ql^2)/8$	0.104
Triangular load q	$(ql^2)/15.6$	0.102
UDL q with end moments M_A, M_B	M_A, M_C, M_B	$k + 0.104\left(1 - \dfrac{\beta}{10}\right)$ $\beta = \dfrac{M_A + M_B}{M_C}$
Cantilever with point load W at al	Triangular, Wal	End deflection $= \dfrac{a(3-a)}{6}$ Load at end $k = 0.333$
Cantilever with UDL q over al	$(qa^2l^2)/2$	$\dfrac{a(4-a)}{12}$ If $a = l$, $k = 0.25$
Point load with end moments M_A, M_B	M_A, M_C, M_B	$k = 0.083\left(1 - \dfrac{\beta}{4}\right)$ $\beta = \dfrac{M_A + M_B}{M_C}$
Two point loads at al from each support (trapezoidal loading)	$[(Wl^2)/24](3 - 4a^2)$	$\dfrac{1}{80} \dfrac{(5 - 4a^2)^2}{3 - 4a^2}$

the maximum sagging moment, even though the hogging moment over the support may be bigger.

An alternative simplified approach is to calculate the deflections on the basis of the uncracked and the fully cracked section and then use the distribution coefficient to combine them. The result would be the same, although the shrinkage curvatures would have had to be converted to deflections.

8.4.5.5. ACCURACY OF DEFLECTION CALCULATIONS.
Many sources of uncertainty will influence the reliability of deflection calculations, including the

- actual level of loading relative to design loading
- variability of tensile strength of concrete
- variability of elastic modulus of concrete
- variability of creep and shrinkage
- behaviour of a cracked tension zone
- stiffening effect of non-structural elements and finishes
- temperature effects
- age when loading is applied and load history.

Many of these factors can have very large effects on the actual deflection. Hence, while in the laboratory where all the above effects are known or are measurable

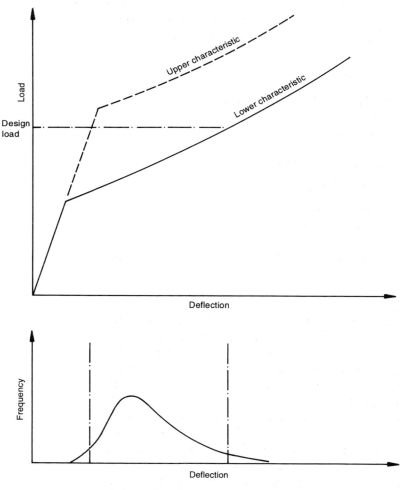

Fig. 8.44. *Variability of deflection in a lightly reinforced member*

an accuracy of around 20% may be achievable, the accuracy in practice is likely to be far lower.

A factor of particular significance in the assessment of the deflection of slabs is the possible variation in tensile strength. This arises because the design load for deflections is commonly close to the cracking moment. The situation is shown in Fig. 8.44, which indicates the range of possible load–deflection curves that might be expected to occur as a result of variations in the cracking moment. It will be seen that this effect can lead to a very wide dispersion in the possible deflections. Other factors can have an equally dramatic effect on the precision of calculations.

Because of this wide dispersion, it may be worth carrying out analyses to establish the possible range of deflections likely to occur, rather than just a single determination, in order to gain some idea of the reliability of the calculation.

8.4.6. Simplified approach to checking deflections

As has been seen, the calculation of deflections is a tedious operation and the results will be of limited reliability. Most codes therefore do not require explicit calculations to be carried out to check deflections except in special cases. The commonest method for controlling deflections is to use limiting ratios of span to effective depth, and this is the approach adopted by EC2.

The basic logic of the use of span/depth ratios to control deflections is straightforward. Consider a simply supported elastic beam supporting a uniformly distributed load of n per unit length, and assume that the maximum permissible stress in the material is f. The maximum moment that the critical section can withstand is thus given by

$$M = fz = nl^2/8$$

where z is the section modulus, which for a rectangular section is $bh^2/6$.

The deflection of the beam is given by

$$a = 5nl^4/384EI$$

Substituting for n from the previous equation gives

$$a = 40fzl^4/384EIl^2$$

Rearranging and substituting βh for I/z gives

$$a/l = 40f/384\beta E(l/h)$$

Since, for a given section shape and material, $40f/384\beta E$ is a constant, this can be rewritten as

$$(a/l) = k(l/h)$$

Thus, for an elastic material, limiting the span/depth ratio will limit the ratio of the deflection to the span. This is an entirely rigorous way of controlling the deflection in these circumstances provided that the limits to deflection are expressed as fractions of the span. Reinforced concrete does not strictly fit the assumptions on which this analysis is based: however, the differences are not as large as they might at first appear. There are two problems to consider: the permissible stress f and the basic stiffness properties of the section, which may be considered to be equivalent to βE. Since the use of span/depth ratios is limited to the consideration of beams and slabs where the tension reinforcement will yield at ultimate, this effectively defines a limiting steel stress under service conditions that can be considered to be equivalent to f. Since the stress limit is applied at the level of the reinforcement rather than at the extreme fibre of the beam, it is more logical to use ratios of span to effective depth than of span to overall depth. Rules can be drafted either in terms of the steel stress under service conditions or in terms of the characteristic strength, effectively assuming a constant relationship between ultimate and service stress. The second of these options is the simpler for the user, and is the approach adopted in EC2. The question of the stiffness properties is more complex, but can be dealt with relatively easily by assuming that, for a given section shape, βE will vary as a function of the reinforcement ratio.

One way of presenting such a system is to provide a set of basic span/effective depth ratios as a function of support conditions (i.e. simply supported, continuous, cantilever, etc.) and a table of modification factors that are a function of steel percentage and steel characteristic strength. A further factor will be needed to correct for section shape. This is the approach that has been used in UK codes for the past 20 years.

In order to develop suitable modification factors for use in EC2, an extensive parameter study was carried out. Deflections were calculated using the calculation method described above for an extensive range of material properties, section geometries and ratios of permanent loads to characteristic loads. For each combination of variables a relationship was obtained between span/effective depth ratio and reinforcement ratio. In these calculations, it was assumed that the characteristic loading was $1/1.43$ times the ultimate loading and that the quasi-permanent load was 0.75 times the characteristic load. The stress in the reinforcement, assuming a cracked section, is thus given approximately by

$$f_{s.qp} = 0.75/(1.43 \times 1.15)f_{yk} = 0.46f_{yk}$$

The result of this exercise was the set of curves shown in Fig. 8.45, combined with the basic ratios given in Table 8.8.

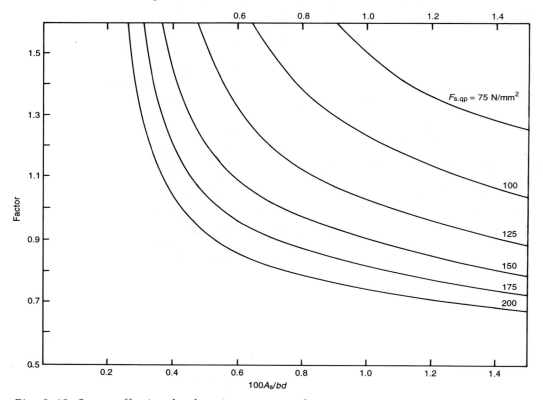

Fig. 8.45. *Span–effective depth ratio correction factors*

Table 8.8. Basic span/effective depth ratios

Support condition	Basic l/d
Simply supported beam, one- or two-way simply supported slabs	25
End span of continuous beam or one-way continuous slab or two-way continuous slab continuous over one long side	32
Interior span of beam or one- or two-way spanning slab	35
Flat slab	30
Cantilever	10

The values given in Table 8.8 should be multiplied by 0·8 for flanged beams where the ratio of the flange width to the rib width exceeds 3. For values of this ratio between 1 and 3, linear interpolation may be used between 1·0 and 0·8.

This approach was considered to be too complex for EC2, so a simplified table was produced giving span effective depths for all the support conditions for two reinforcement ratios: 0·5% and 1·5%. Furthermore, the values were expressed as a function of the characteristic reinforcement stress rather than the stress under the quasi-permanent load. These are the columns of values marked 'concrete lightly stressed' and 'concrete highly stressed' respectively in Table 4.14.

Example 8.5
Check whether or not the internal panel of a slab with a span of 6 m, an effective depth of 200 mm and a reinforcement percentage of 0·5 will be satisfactory from the point of view of deflections. The characteristic strength of the reinforcement is 400 N/mm².

Clause 4.4.3.2(5)a states that slabs can normally be assumed to be 'lightly stressed', hence, from *Table 4.14*, the permissible span/effective depth is 35. The The actual span/effective depth ratio is 6000/200 = 30. This is less than the permissible value, therefore the slab is satisfactory.

Alternatively, Fig. 8.45 could be used by assuming that the stress under the quasi-permanent load is $0·46 \times 400 = 184$ N/mm². It will be seen that Fig. 8.45 gives a factor of 1·0 for 0·5% of tension steel and this stress. The permissible span/effective depth ratio is this $1·0 \times 35 = 35$; the same answer as given by *Table 4.14*. If, however, the reinforcement ratio had been 0·35% instead of 0·5%, Fig. 8.45 would have given a factor of 1·25 instead of 1·0, and hence the allowable ratio would have been 43·8. It will be seen that EC2 results are very conservative compared with the results from the parameter study for reinforcement percentages below 0·5%, which are very common for slabs.

Clause 4.4.3.2(5)a

CHAPTER 9

Durability

9.1. General

The Construction Products Directive defines certain Essential Requirements relating to fitness for purpose, mechanical resistance and stability, and various other factors. These requirements 'must, subject to normal maintenance, be satisfied for an economically reasonable working life'. A structure having adequate durability is one that satisfies this last stipulation. This chapter is concerned with the rules given in EC2 and associated documents to ensure that adequate durability is achieved. Design for durability is not covered fully by EC2, and reference needs to be made to two other documents.[15,16]

In EC2, durability is mainly covered in *section 4.1*, but parts of *clause 4.4.2* are also relevant.

Section 4.1
Clause 4.4.2

Additional information on the background to the provisions can be found in the Supporting Document on durability (date and mode of publication not yet known). Further detailed information on many aspects of design for durability can be found in *section 9.1*.

Section 9.1

9.2. Background

9.2.1. Historical perspective

Until about 15 years ago, durability was not seen as a serious issue for concrete. While all codes gave minimum covers for protection of the reinforcement, it was rarely felt necessary to do more than this. For example, durability is not mentioned explicitly in the contents list of the CEB 1978 Model Code, which formed the base document for the drafting of EC2. During the 1970s, however, durability problems arose in many countries, and this has resulted in a complete change in attitude to the design and construction of concrete structures. The more notable problems include the following.

(a) Very serious deterioration of bridge decks in the USA due to corrosion of the reinforcement which was contaminated by de-icing salts.
(b) Major deterioration problems in the Middle East due to chloride-induced corrosion in a particularly aggressive environment.
(c) Major problems in the UK in the 1970s due to reductions in strength of high alumina cement with time. More recently, this has also become a major problem in Spain.
(d) Severe cracking in structures in many countries resulting from alkali–silica reaction. The number of affected structures is not high, but the problem has generated a large amount of publicity.

(e) Problems resulting from corrosion initiated by de-icing salts on bridges in almost all countries where these are used in significant quantities in winter.

As a result of these problems, durability and, in particular, corrosion of reinforcement were probably the major areas of research in the structural field during the 1980s. Inevitably, the drafters of codes of practice also found it essential to treat the subject much more fully. EC2 has followed this trend.

9.2.2. Common mechanisms leading to the deterioration of concrete structures

This section enumerates the main deterioration mechanisms that may need to be considered in design, and gives a brief description of the phenomena involved and the possible methods for dealing with them.

9.2.2.1. CORROSION OF REINFORCEMENT OR PRESTRESSING TENDONS. In normal circumstances, the highly alkaline nature of concrete protects steel embedded within it. Except under the circumstances discussed below, the pH value of the pore solution in concrete is in the region of 12–14. Steel will not generally corrode in uncontaminated concrete until the pH drops below 10. The protection is afforded by the formation of a very thin, coherent layer of iron oxide over the surface of the bar in alkaline conditions. Steel protected in this way is described as being in a passive state. Two mechanisms can lead to the destruction of this passive state: carbonation of the concrete and the action of chlorides.

9.2.2.1.1. Carbonation of the concrete. This is a reaction between carbon dioxide in the atmosphere and the alkalis in the cement matrix. This process starts at the surface and, with time, penetrates slowly into the concrete. The rate of penetration of carbonation into the concrete depends on the environment and the quality of the concrete. The rate is highest where the relative humidity is in the range 50%–60%. It is lower at higher humidities, and effectively zero at 100%. Good-quality concrete carbonates more slowly than poor-quality concrete. In this instance, the rate depends on the rate at which carbon dioxide can diffuse into the concrete. This will decrease with decreasing water/cement ratio, and hence with increasing strength. The effect of carbonation is to reduce the alkalinity of the concrete to a level where the natural protection is lost: corrosion may then occur if the concrete immediately surrounding the reinforcement is carbonated.

9.2.2.1.2. The presence of chlorides in the concrete. Chlorides have the capacity to destroy the passivity of steel even where the alkalinity remains high. This usually occurs locally, giving rise to 'pitting corrosion'. Chlorides may get to the concrete from various sources, the commonest being sea water in marine environments, de-icing chemicals on roads and additives such as calcium chloride, which was used extensively in the past. The rate at which chlorides penetrate into concrete depends on the rate of application of chlorides to the concrete surface and, as with carbonation, on the quality of the concrete.

9.2.2.1.3. Active corrosion. Once the passivity of the steel has been destroyed, corrosion can occur if there is (*a*) sufficient moisture and (*b*) sufficient oxygen. It is found that these two requirements can act against each other, since if the concrete is wet oxygen cannot penetrate, and if it is dry so that there is plentiful oxygen, there is insufficient moisture for the reaction to progress. As a result, the greatest risk of corrosion is in members subjected to wetting and drying.

The normal way to design against corrosion is to ensure that there is an adequate cover to the reinforcement and that the concrete in the cover region is of a high quality and is well cured. In particularly aggressive environments, however, other, more expensive measures may be taken, including

(a) the use of reinforcement coated with epoxy or similar
(b) the use of stainless steel reinforcement
(c) the application of surface coatings to the concrete to inhibit the ingress of chlorides or carbon dioxide — to be successful for long periods of time, such coatings would have to be meticulously maintained
(d) the application of cathodic protection to the structure.

One major factor in the avoidance of corrosion problems is the form of the structure. Areas of exposed concrete on which water can stand or which water can drain across are particularly at risk.

9.2.2.2. FROST ATTACK.

If saturated concrete is subjected to frequent freezing and thawing, the expansive effects of ice will disrupt the concrete. The usual manifestations of frost damage are surface spalling or the formation of systems of closely spaced surface cracks. These cracks can be seen as the precursors of spalling. Concrete that is not close to being saturated is not at risk from frost, as the expansion that occurs on freezing can be accommodated in the non-water-filled pores. Frost damage can be avoided by

(a) protecting the concrete from saturation
(b) using an air-entrained concrete mix — the small bubbles within the matrix can provide pressure relief
(c) using high-strength concrete — concrete of strength 45 N/mm^2 or more is generally immune to frost damage.

9.2.2.3. ALKALI–AGGREGATE REACTIONS.

There are two basic forms of reaction that occasionally occur and can damage concrete: alkali–silica reaction and alkali–carbonate reaction. The alkali–silica reaction is the more common. It is a reaction between the alkalis in the cement and certain forms of silica which results in the formation of a hygroscopic silica gel. This gel takes up water and expands, causing cracking. Identification of alkali–silica reaction as the cause of damage is not straightforward. The first stage is to establish that silica gel has formed within the matrix. This can be done only by taking thin sections of the concrete and examining them under a microscope to establish the existence of gel-filled micro-cracks. Establishing the presence of gel is not sufficient as proof that alkali–silica reaction is the cause of the observed damage, as limited amounts of gel can frequently be present in concrete without any deleterious consequences. There must be evidence of large quantities of gel. Externally, the effect of these reactions is the formation of cracks, which may be large (several millimetres wide is not uncommon). In relatively unstressed and unreinforced concrete, these cracks can form a random 'map' pattern. In other cases they will tend to form parallel to the direction of compressive stress or reinforcement. Although large, the cracks are usually not deep, extending only 50–70 mm into the section. Their effect on structural performance is not as great as might be imagined from looking at the cracks. A reduction in the compressive and tensile strength of the concrete occurs, but this is commonly not more than about 20%–30%.

Alkali–silica reaction can be avoided by three methods.

(a) The use of aggregates that experience has shown to perform satisfactorily.

No fully adequate test has yet been devised to assess the potential reactivity of aggregates, so experience remains the only reliable criterion.

(b) The use of cement with a low alkali content. The level that will obviate problems is likely to vary from country to country, depending on the geology. In the UK, a limit of 3 kg of sodium oxide equivalent is considered adequate.

(c) Inhibition of the ingress of water.

9.2.2.4. ATTACK FROM SULPHATES. In the presence of water, sulphate ions can react with the tricalcium aluminate component of the matrix. This reaction causes expansion, leading to cracking and eventual disintegration of the concrete. The commonest source of sulphates is the earth surrounding foundations, but other sources are sometimes significant. Sea water contains significant amounts of sulphate, but the presence of chlorides renders it harmless.

If it is established that there is a risk of sulphate attack, the possible measures to avoid it are

(a) the use of sulphate-resisting cement, i.e. Portland cement with a low tricalcium aluminate content

(b) the use of blended cements incorporating ground granulated blastfurnace slag

(c) for medium levels of risk, the use of additions such as fly ash or pozzolans.

In particularly aggressive situations, it may be necessary to provide an impervious coating to the concrete.

9.2.2.5. ACID ATTACK. Acids attack the calcium compounds in concrete, converting them to soluble salts which can then leach away. The effect of acids is therefore to eat away, or render weak and permeable, the surface of the concrete. Very substantial amounts of acid are required to do serious damage to concrete. Acid rain, for example, will do no more than etch the surface of concrete over any reasonable design life. If the concrete is likely to be exposed to major amounts of acid, for example from some industrial process, the only way to avoid damage is to provide an impermeable coating to the concrete.

9.2.2.6. LEACHING BY SOFT WATER. This process is in effect a mild version of acid attack. Calcium compounds (e.g. calcium carbonate, calcium hydroxide) ate weakly soluble in soft water and can be leached out if the concrete is constantly exposed to running soft water. The process is very slow but, in time, can lead to exposed concrete surfaces taking on an 'exposed aggregate' appearance.

9.2.2.7. ABRASION. Abrasion of concrete surfaces may occur due to trafficking of the concrete or due to sand or gravel suspended in turbulent water.

Resistance to abrasion can be obtained by using higher strength concrete and abrasion-resistant aggregates. Resistance is also markedly improved by good curing of surfaces likely to be exposed to abrasive action.

9.2.3. Relative importance of deterioration mechanisms

Clearly, the relative importance of the various mechanisms will vary from country to country and even from region to region, and no generally applicable ordering of the mechanisms can be made. However, there seems no doubt that the commonest and most serious form of degradation worldwide is corrosion of reinforcement. It can also be stated that, of the two initiating mechanisms for corrosion (carbonation and chlorides), chlorides have led to by far the greater amount of damage.

9.3. Design for durability

9.3.1. General
There are two basic steps in designing for durability, i.e. to

(a) establish the aggressivity of the environment to which the member is exposed: this is analogous to establishing the design loading where the ultimate or serviceability limit states are being considered
(b) select materials and design the structure to be able to resist the environment for a reasonable lifetime.

These steps are now considered in turn.

9.3.2. Definition of aggressivity of the environment
The aggressivity of the environment should, in principle, be defined separately for each degradation mechanism, since the factors acting to promote one form of degradation are not necessarily the same as those promoting another. EC2 and ENV 206 both do this, although the mechanisms are not as clearly separated as they might be. Exposure classes are defined in *Table 4.1*, which is the same as Table 2 in ENV 206. This classifies environments into five basic classes with a number of sub-classes, i.e.

(1) dry environment
(2) humid environment
    (a) without frost
    (b) with frost
(3) humid environment with frost and de-icing salts
(4) sea water environment
    (a) without frost
    (b) with frost
(5) aggressive chemical environment
    (a) slightly aggressive
    (b) moderately aggressive
    (c) highly aggressive.

Each of environments 1–4 is described by examples, which defined aggressivity in relation to corrosion of the reinforcement and frost. Exposure class 5 is not defined in either EC2 or ENV 206. Reference is made to a draft ISO standard, ISO/DIS 9690. This document classifies chemical aggressivity into five levels, A1–A5. Each level is split into three sub-classes, covering aggressiveness from gas, denoted by G, liquids, L, and solids, S. In EC2, exposure class 5(a) corresponds to ISO classes A1G, A1L and A1S; 5(b) to ISO classes A2G, A2L and A2S; and 5(c) to A3G, A3L and A3S.

The most common form of chemical aggressivity is sulphate attack: its aggressivity is defined in terms of the concentration of sulphate ion (SO_4) in the liquid or soil to which the structure is exposed, measured in mg/l or mg/kg as follows

exposure class 5(a)   250–500
exposure class 5(b)   500–1000
exposure class 5(c)   1000–6000

9.3.3. Measures to resist environmental aggressivity

Measures relating to the selection of materials and design of a suitable concrete mix are found in ENV 206; measures relating to the design of the structure, such as cover to the reinforcement, are included in EC2. The design of a durable structure thus requires attention to the rules in both documents. It should be noted, however, that the science of durability is not well advanced, and suitable measures have developed in different countries by trial and error over many years. For this reason, most of the provisions in ENV 206 are prefaced with the proviso that other methods may be used where experience shows them to be satisfactory or where they are required by national standards. The specific durability provisions

Table 9.1. Durability requirements

Parameter	Exposure class								
	1	2(a)	2(b)	3	4(a)	4(b)	5(a)	5(b)	5(c)
Maximum water/cement ratio									
Plain concrete		0·7							
Reinforced concrete	0·65	0·6	0·55	0·5	0·55	0·5	0·55	0·5	0·45
Prestressed concrete	0·6	0·6							
Grade deemed to satisfy w/c limits									
CE 32·5 cement		C25	C30	C35	C30	C35	C30	C35	C40
CE 42·5 cement		C30	C35	C40	C35	C40	C35	C40	C45
Minimum cement content									
Plain concrete	150	200	200				200		
Reinforced concrete	260	280	280	300	300	300	280	300	300
Prestressed concrete	300	300	300				300		
Minimum air content									
32 mm max. aggregate			4	4		4			
16 mm max. aggregate			5	5		5			
8 mm max. aggregate			6	6		6			
Frost-resistant aggregate			Yes	Yes		Yes			
Low permeability			Yes	Yes	Yes	Yes	Yes	Yes	Yes
Minimum cover: mm									
Reinforced concrete slabs, concrete <C40/50	15	15	20	35	35	35	20	25	35
Reinforced concrete slabs, stronger concrete	15	15	15	30	30	30	15	20	35
Other, concrete <C40/50	15	20	25	40	40	40	25	30	40
Other, stronger concrete	15	15	20	35	35	35	20	25	40
Prestressed concrete slabs, concrete <C40/50	25	25	30	45	45	45	30	35	45
Prestressed concrete slabs, stronger concrete	25	25	25	40	40	40	25	30	45
Other, concrete <C40/50	25	30	35	50	50	50	35	40	50
Other, stronger concrete	25	25	30	45	45	45	30	35	50

Sulphate-resisting cement should be used in exposure class 5 where the sulphate ion content exceeds 500 mg/kg in water or 3000 kg/mg in soil.

In exposure class 5(c), a protective barrier should be used to prevent direct contact between the aggressive media and the concrete.

The curing times given in clause 10.6.3 and Table 12 of ENV 206 should also be complied with if adequate durability is to be assured.

in EC2 are mostly 'boxed' for the same reason. The main provisions of EC2 and ENV 206 are summarized in Table 9.1. It should be checked that they are applicable in the particular country where the structure is to be built.

Table 9.1 combines the provisions of Tables 3 and 20 in ENV 206 and *Table 4.2* in EC2.

CHAPTER 10

Detailing

10.1. General
The main guidance to detailing is in *Chapter 5* of this Code. Cover requirements are in *section 4.1*, under 'durability'. Most of the provisions regarding anchorages and laps were reputedly deduced from tests carried out in many countries, principally in Austria, France, Germany, Sweden and the USA. It is understood that the global safety factor $(\gamma_m \cdot \gamma_f)$ against the 5% fractile of tests results is $2 \cdot 1$. As $\gamma_m = 1 \cdot 5$ for concrete this gives a value of $\gamma_f = 1 \cdot 4$ for loads, which the committee clearly considers acceptable.

Chapter 5
section 4.1

Anchorage and lap requirements are checked at the ultimate limit stage. The general detailing provisions are deemed to ensure satisfactory behaviour of the structure at serviceability conditions.

EC2 covers (*a*) high-bond bars, (*b*) plain bars, (*c*) welded mesh using high-bond wires and (*d*) different types of structural member. Bars may be used in bundles with certain limitations.

The detailing requirements are mainly governed by bond-related phenomena, which are significantly influenced by

(*a*) the surface characteristics of the bars (plain, ribbed)
(*b*) the shape of the bars (straight, with hooks or bends)
(*c*) the presence of welded transverse bars
(*d*) the confinement offered by concrete (mainly controlled by the size of the concrete cover in relation to the bar diameter)
(*e*) the confinement offered by non-welded transverse reinforcement (such as links)
(*f*) the confinement offered by transverse pressure.

The rules governing detailing allow for the above factors. Particular emphasis is placed on the need for adequate concrete cover and transverse reinforcement to cater for tensile stresses in concrete in regions of high bond stresses.

Bond stresses for plain bars are related to the cylinder strength of concrete f_{ck}; those for high-bond bars are a function of the tensile strength of concrete f_{ctk}.

The guidance for detailing of different types of member includes requirements for minimum areas of reinforcement. This is stipulated in order to (*a*) prevent a brittle failure, (*b*) prevent wide cracks, and (*c*) resist stresses arising from temperature effects, shrinkage and other restrained actions.

10.2. Discussion of the general requirements
In this section, the main features of the detailing requirements are arranged in a practical order and discussed.

10.2.1. Cover to bar reinforcement
See Table 10.1.

10.2.2. Spacing of bars
See Fig. 10.1.

Clause 5.2.1.1

10.2.3. Minimum diameters of bends

Clause 5.2.1.2

See Tables 10.2 and 10.3. Although this is not stated explicitly, the diameters of bends specified in Table 10.3 relate to fully stressed bars; linear interpolation is permissible for other stress levels.

10.2.4. Bond

10.2.4.1. BOND CONDITIONS. Two bond conditions (good and poor) are defined. These take note of the likely quality of concrete as cast, and are illustrated

Table 10.1. Cover requirements to bar reinforcement

Minimum cover: mm	Exposure class								
	1	2(a)	2(b)	3	4(a)	4(b)	5(a)	5(b)	5(c)
	15	20	25	40	40	40	25	30	40

For concrete cast directly against the ground, minimum cover = 75 mm.
For concrete cast against prepared ground, minimum cover = 40 mm.
The minimum cover should be increased by an allowance for tolerance which is normally in the range 5–10 mm.
The cover should never be less than ϕ or ϕ_n ($\not> 40$ mm), or ($\phi + 5$ mm) or ($\phi_n + 5$ mm) where aggregates larger than 32 mm are used (ϕ is the diameter of the bar; ϕ_n is the equivalent diameter for a bundle of bars).

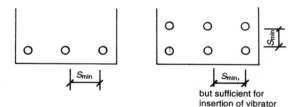

Fig. 10.1. Spacing of bars

Table 10.2. Minimum diameters for hooks, bends and loops

	Bar diameter	
	$\phi < 20$ mm	$\phi \geq 20$ mm
Plain bars $f_{yk} = 250$	$2 \cdot 5\phi$	5ϕ
High-bond bars $f_{yk} = 460$	4ϕ	7ϕ

Table 10.3. Minimum diameters for bent-up bars or other curved bars

	Minimum cover perpendicular to the plane of curvature		
	>100 mm and >7ϕ	>50 mm and >3ϕ	≤50 mm and ≤3ϕ
Plain bars $f_{yk} = 250$	10ϕ	10ϕ	15ϕ
High-bond bars $f_{yk} = 460$	10ϕ	15ϕ	20ϕ

DETAILING

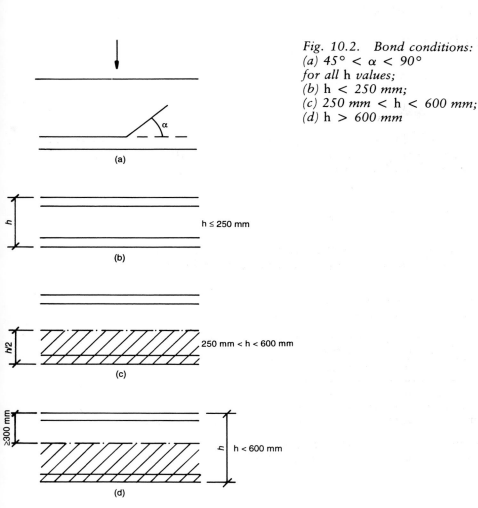

Fig. 10.2. Bond conditions:
(a) $45° < \alpha < 90°$ for all h values;
(b) $h < 250\ mm$;
(c) $250\ mm < h < 600\ mm$;
(d) $h > 600\ mm$

in Fig. 10.2. Fig. 10.2(a) and (b) shows good bond conditions for all bars. Fig. 10.2(c) and (d) shows good bond conditions for bars in hatched zones, and poor bond conditions for bars in zones that are not hatched.

10.2.4.2. ULTIMATE BOND STRESS. See Table 10.4. Where bond conditions are 'poor', the values in Table 10.4 should be multiplied by 0·7. Where a mean

Clause 5.2.2.2

Table 10.4. Values of ultimate bond stress

f_{ck}: N/mm²	Ultimate bond stress f_{bd}: N/mm²	
	Plain bars	High-bond bars $\phi \leq 32$
12	0·9	1·6
16	1·0	2·0
20	1·1	2·3
25	1·2	2·7
30	1·3	3·0
35	1·4	3·4
40	1·5	3·7
45	1·6	4·0
50	1·7	4·3

pressure p (N/mm^2) exists transverse to the plane of splitting, the values should be multiplied by $1/(1 - 0.04p) \leq 1.4$.

For requirements relating to bar diameters > 32 mm, see section 10.2.7 below.

10.2.5. Anchorage

10.2.5.1. BASIC ANCHORAGE LENGTH l_b. If the axial force in the bar is equated to the bond forces developed on a length l_b, the result is $l_b = (\phi/4)(f_{yd}/f_{bd})$.

10.2.5.2. METHODS OF ANCHORAGE. See Fig. 10.3.

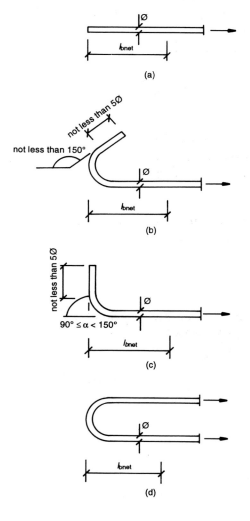

Fig. 10.3. Methods of anchorage: (a) straight bar (not permitted for smooth bars > 8 mm diameter); (b) hook; (c) bend; (d) loop; (e) welded transverse bar

10.2.5.3. ANCHORAGE LENGTH REQUIRED $l_{b,net}$.

$$l_{b,net} = \alpha_a l_b (A_{s,req}/A_{s,prov}) \not< l_{b,min}$$

Clause 5.2.3.4
clause 5.2.3.4

where α_a equals $1 \cdot 0$ for straight bars, and equals $0 \cdot 7$ for curved bars in tension if the concrete cover perpendicular to the plane of curvature is at least 3ϕ; l_b is the basic anchorage length; $l_{b,min}$ is the minimum anchorage length $= 0 \cdot 6 l_b$ for tension, $= 0 \cdot 3 l_b$ for compression, $\not< 10\phi$ or $\not< 100$ mm).

If welded transverse bars are present in the anchorage, the above expression for $l_{b,net}$ may be multiplied by $0 \cdot 7$.

10.2.5.4. TRANSVERSE REINFORCEMENT.

At anchorage, tensile stresses are induced in concrete which tend to split the concrete cover. Lateral reinforcement should be provided to cater for these lateral tensile stresses (Fig. 10.4).

Clause 5.2.3.3.

Transverse reinforcement should be provided for

(a) anchorage in tension, if no compression is caused by support reactions
(b) all anchorages in compression.

In tension anchorages, the transverse reinforcement should be evenly distributed along the anchorage length, with at least one bar placed in the region of a hook, bend or loop.

In compression anchorages, the transverse reinforcement should surround the bars and be concentrated at the end of the anchorage, as some of the forces will be transferred by the end of the bar (pin effect) and this in turn will result in bursting forces.

10.2.5.5. ANCHORAGE OF LINKS.

Links and shear reinforcement may be anchored using one of the methods shown in Fig. 10.5. In Fig. 10.5(c) and (d), the transverse bars are welded.

Clause 5.2.5

10.2.6. Splices (laps)

10.2.6.1. SPACES BETWEEN ADJACENT LAPS.

Laps between bars should be staggered and should not be located at sections of high stress. Spaces between lapped bars should comply with the requirements shown in Fig. 10.6.

Clause 5.2.4.1.1

10.2.6.2. LAP LENGTHS l_s.

$$l_s = \alpha_1 l_{b,net} \not< l_{s,min}$$

Clause 5.2.4.1

where

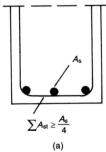
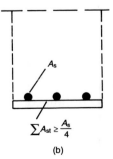

(a) (b)

Fig. 10.4. *Lateral reinforcement: (a) beam; (b) slab* (A_{st} = *area of one bar of the transverse reinforcement;* A_s = *area of one anchored bar*)

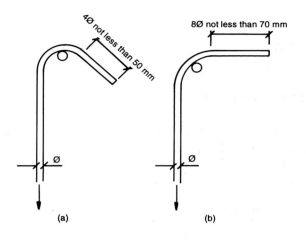

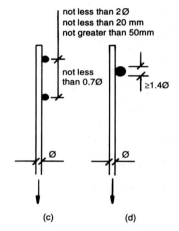

Fig. 10.5. Anchorage of links

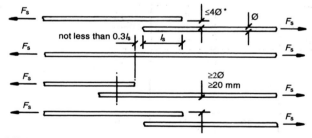

* If > 4 Ø, the lap length shall be increased by the amount by which the clear gap exceeds 4Ø

Fig. 10.6. Spaces between lapped bars

$\alpha_1 = 1 \cdot 0$ for compression laps and for tension laps where

(a) less than 30% of the bars at a section are lapped
(b) the clear distance between adjacent lapped bars $> 10\phi$ and the cover $> 5\phi$ when ϕ is the diameter of the bar.

$\alpha_1 = 1 \cdot 4$ for tension laps where either

(a) 30% or more of the bars at a section are lapped or
(b) the clear distance between adjacent lapped bars $< 10\phi$ or the cover $< 5\phi$.

$\alpha_1 = 2 \cdot 0$ for tension laps where both (a) and (b) for $\alpha_1 = 1 \cdot 4$ above are satisfied.

$$l_{s,\min} = 0 \cdot 3 \alpha_a \alpha_1 l_b \not< 15\phi \not< 200 \text{ mm}$$

Table 10.5. Anchorage and lap lengths as multiples of bar size: deformed bars type 2 (to BS4449), $f_{yk} = 460$ N/mm²

Concrete strength: N/mm²	f_{ck}	20	25	30	35	40
	f_{cu}	25	30	37	45	50
Anchorage: straight bars, compression and tension		44	37	34	30	27
Anchorage: curved bars*, tension		31	26	24	21	19
Laps: compression, tension†		44	37	34	30	27
Laps: tension‡		62	52	48	42	38
Laps: tension§		88	74	68	60	54

The values in the table apply to (a) good bond conditions (see section 10.2.4.1) and (b) bar size ≤ 32.
  For poor bond conditions (see section 10.2.4.1), the table values should be divided by 0·7.
  For bar size > 32 the values should be divided by $(132 - \phi)/100$, where ϕ is the bar diameter in mm.
* In the anchorage region, cover perpendicular to the plane of curvature should be at least 3ϕ.
† The percentage of bars lapped at the section <30%, clear spacing between bars ≥10ϕ and side cover to the outer bar ≥5ϕ.
‡ The percentage of bars lapped at the section >30%, *or* clear spacing between bars <10ϕ, or side cover to the outer bar <5ϕ.
§ The percentage of bars lapped at the section >30% *and* clear spacing between bars <10ϕ or side cover to the outer bar <5ϕ.

Table 10.6. Anchorage and lap lengths as multiples of bar size: plain bars, $f_{yk} = 250$ N/mm²

Concrete strength: N/mm²	f_{ck}	20	25	30	35	40
	f_{cu}	25	30	37	45	50
Anchorage: straight bars, compression and tension (not applicable to bar diameter > 8 mm)		50	46	41	39	37
Anchorage: curved bars*, tension		35	32	29	27	26
Laps: compression, tension†		50	46	41	39	37
Laps: tension‡		70	64	60	56	52
Laps: tension§		100	92	84	78	74

The values in the table apply to good bond conditions (see section 10.2.4.1).
  For poor bond conditions (see section 10.2.4.1), the table values should be divided by 0·7.
* In the anchorage region, cover perpendicular to the plane of curvature should be at least 3ϕ.
† The bars lapped at the section <30%, clear spacing between bars ≥10ϕ and side cover to the outer bar ≥5ϕ (from NAD).
‡ The bars lapped at the section >30%, *or* clear spacing between bars <10ϕ, or side cover to the outer bar <5ϕ.
§ The bars lapped at the section >30% *and* clear spacing between bars <10ϕ or side cover to the outer bar <5ϕ.

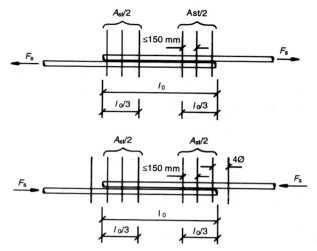

Fig. 10.7. Placing of transverse reinforcement

where α_a and l_b are as defined in section 10.2.5.3 and 1 above, and α_1 is as defined above.

Anchorage and lap lengths required in different grades of concrete are given in Tables 10.5 and 10.6 for high-bond and plain bars.

10.2.6.3. TRANSVERSE REINFORCEMENT AT LAPPED JOINTS.
As at anchorages, tensile stresses are induced in concrete at lapped joints and these stresses tend to split the concrete cover. Lateral reinforcement should be provided to resist these stresses. Failure of splices without transverse reinforcement is sudden and complete, whereas those with transverse reinforcement tend to exhibit a less brittle failure and also possess residual strength beyond the maximum load.

No special reinforcement is required when the diameter of the lapped bars is less than 16 mm, or the lapped bars in any section are less than 20%. Under these conditions the minimum reinforcement is considered adequate to cope with the tensile stresses generated at laps.

Clause 5.2.4.1.2

If the diameter of the lapped bars is greater than 16 mm, transverse reinforcement should be placed between the longitudinal reinforcement and the concrete surface as shown in Fig. 10.7.

Where the clear distance between adjacent lapped bars $\leq 10\phi$, the transverse reinforcement should be in the form of links in beams.

10.2.7. Bars with $\phi > 32$ mm

10.2.7.1. GENERAL.
The minimum depth of the element should not be less than 15ϕ.

For crack control, surface reinforcement may be used or crack width should be calculated and justified.

Concrete cover should be greater than ϕ. The clear distance (horizontal and vertical) between bars should not be less than ϕ or the maximum aggregate size + 5 mm.

10.2.7.2. BOND.
The values of ultimate bond stress should be multiplied by $((132-\phi)/100)\phi$ (in mm).

Clause 5.2.6.2.

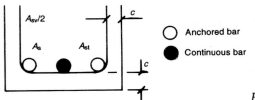

Fig. 10.8. Additional transverse reinforcement

10.2.7.3. ANCHORAGE.

(a) Bars should be anchored as straight bars or by means of mechanical devices. They should not be anchored in tension zones. *Clause 5.2.6.3*

(b) Lapped joints should not be used and mechanical devices (e.g. couplers) should be considered.

(c) In the absence of transverse compression additional transverse reinforcement should be provided as shown in Fig. 10.8.

$$A_{st} = n_1 0.25 A_s$$
$$A_{sv} = n_2 0.25 A_s$$

where A_s is the cross-sectional area of the anchored bar, n_1 is the number of layers with anchored bars in the same section, and n_2 is the number of bars anchored in each layer.

(d) The additional transverse bars should be distributed evenly in the anchorage zone with their spacing not exceeding 5ϕ.

10.2.8. Welded mesh
Clause 5.2.4.2

10.2.8.1. MINIMUM DIAMETERS OF MANDRELS.
Clause 5.2.1.2

The diameter depends on whether the welded cross wires are inside or outside the bends and on their location with respect to the tangent point of the bend (see Fig. 10.9).

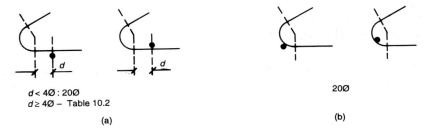

Fig. 10.9. Minimum diameter of the mandrel: (a) welds outside bends; (b) welds inside bends

10.2.8.2. LAPS FOR WELDED MESH FABRICS MADE OF HIGH-BOND WIRES.
Clause 5.2.4.2

10.2.8.2.1. General. Mesh reinforcement may be lapped by (a) intermeshing (the lapped wires occuring in one plane) or (b) layering (the lapped wires occurring in two planes separated by the crosswires).

When intermeshing is used in one direction, the wires at right angles will automatically be layered (see Fig. 10.10).

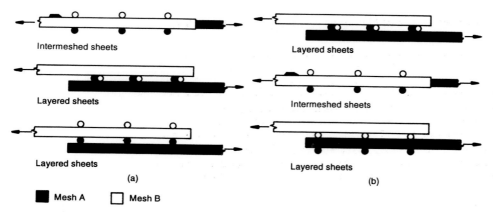

Fig. 10.10. *Layering of wires: (a) main reinforcement; (b) transverse reinforcement*

Table 10.7. Amount of main reinforcement that may be lapped

A_s/s	Interior mesh	Exterior mesh
≤ 1200 mm²/m	100%	100%
> 1200 mm²/m	60%	Laps not allowed

EC2 does not provide guidance for lapping by intermeshing, which is the most efficient method. There is no technical reason not to use the EC2 provisions for intermeshing.

10.2.8.2.2. Location of laps (main reinforcement). Laps should be in zones where the effects of actions under the rare combination of loads are not more than 80% of the design strength of the section.

The amount of main reinforcement that may be lapped in any one section depends on the specific section area of the mesh, denoted by A_s/s (i.e. area of reinforcement per unit width), and whether the mesh is an interior or exterior mesh in a multiple layer mesh. Table 10.7 may be used.

10.2.8.2.3. Lap length.

$$\text{Lap length } l_o = \alpha_2 l_b (A_{s,req}/A_{s,prov})$$
$$\not< l_{o,min}$$

where

$$\alpha_2 = 0.4 + [(A_s/s)/800]$$
$$\not< 1 \text{ and } \not> 2$$

l_b is the basic anchorage length (see section 10.2.5.1.).

$$l_{o,min} = 0.3 \, \alpha_2 \, l_b$$
$$\not< 200 \text{ mm}$$
$$\not< S_t, \text{ the spacing of transverse welded bars}$$

The lap lengths required may be expressed as multiples of the diameter of the main reinforcement bars, as in Table 10.8.

DETAILING

Table 10.8. Lap lengths l_b for weld mesh made of high-bond wires (f_{yk} = 460 N/mm²) as multiples of main wire size

Concrete strength f_{ck}: N/mm²	20	25	30	35	40
Basic lap length*	50	43	38	34	31

The values in the table apply to (a) good bond conditions (see section 10.2.4.1) and (b) bar size ≤32.
For poor bond conditions (see section 10.2.4.1), the table values should be divided by 0·7.
For bar size >32 the values should be divided by [(132 − φ)/100], where φ is the bar diameter in mm.

* The basic lap length applies to mesh with A_s/s up to 480 mm²/m. For mesh with A_s/s between 480 and 1280 mm²/m, the basic lap length should be multiplied by α_2, obtained by linear interpolation between the following values: for A_s/s = 480 mm²/m, α_2 = 1·00; for A_s/s ≥ 1280 mm²/m, α_2 = 2·00.

Table 10.9. Minimum lap length requirements

Diameter of transverse bars	Minimum lap length
φ ≤ 6 mm	150 mm
6 mm < φ < 8·5 mm	250 mm
8·5 mm < φ ≤ 12 mm	350 mm

10.2.8.2.4. Laps for transverse distribution reinforcement. All transverse bars may be lapped at the same location.

The lap length should be at least equal to S_l (the spacing, of the longitudinal wires) or the values given in Table 10.9.

10.2.9. Welded mesh using smooth wires

EC2 does not provide direct guidance on this, but refers to national codes. In the UK, BS8110 provides guidance for such a mesh. Table 10.10 may be used to determine the lap length.

Table 10.10. Anchorage and lap lengths as multiples of bar size: smooth wire fabric, f_{yk} = 460 kN/mm²

Concrete strength: N/mm²	f_{ck}	20	25	30	35	40
	f_{cu}	25	30	37	45	50
Straight anchorage: compression		26	24	22	20	19
Straight anchorage: tension		33	30	27	25	23
Laps: compression, tension*		33	30	27	25	23
Laps: tension†		46	42	38	34	33
Laps: tension‡		66	60	54	49	47

The values in the table apply to (a) good bond conditions (see section 10.2.4.1) and (b) fabric defined in BS4449, BS4461 and BS4482.
For poor bond conditions (see section 10.2.4.1), the table values should be divided by 0·7.
Bond stresses are based on BS8110:1985, but modified to allow for γ_c = 1·5.
The values apply provided: the fabric is welded in a shear-resistant manner complying with BS4483, and the number of welded intersections within the anchorage is at least equal to 4 × (A_{sreq}/A_{sprov}). If the latter condition is not satisfied, values appropriate to the individual bars/wires should be used.
* The bars lapped at the section <30%, clear spacing between bars ≥10φ and side cover to the outer bar ≥5φ.
† The bars lapped at the section >30%, or clear spacing between bars >10φ, or side cover to the outer bar >5φ.
‡ The bars lapped at the section >30% and clear spacing between bars >10φ or side cover to the outer bar >5φ.

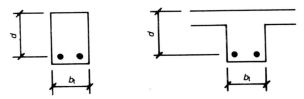

Fig. 10.11. Minimum area of longitudinal reinforcement

10.2.10. Beams

10.2.10.1. LONGITUDINAL REINFORCEMENT.

10.2.10.1.1. Minimum area.
Minimum area $A_{st,min} \not< (0 \cdot 6 b_t d / f_{yk}) \not< 0 \cdot 0015 \, b_t d$, where f_{yk} is the characteristic yield stress of reinforcement (Fig. 10.11).

Clause 5.4.2.1.1(1)

At supports in monolithic construction where simple supports are assumed in the design (Fig. 10.12), A_{st}(support) $\not< (1/4) \, A_{st}$(span).

Clause 5.4.2.1.2(1)

10.2.10.1.2. Maximum area. Maximum area $A_{st,max}$ or $A_{sc,max} \not> 0 \cdot 04 A_c$, where A_c is the cross-sectional area of concrete (Fig. 10.13).

Clause 5.4.2.1(2)

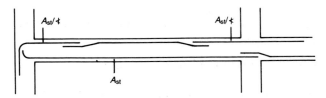

Fig. 10.12. Longitudinal reinforcement at supports in monolithic construction

Fig. 10.13. Maximum area of longitudinal reinforcement

10.2.10.1.3. Distribution of support reinforcement at internal supports of continuous flanged beams.

Clause 5.4.2.1.2(2)

Total support reinforcement A_{st} may be distributed approximately equally between the internal and external parts of the flange (Fig. 10.14).

10.2.10.2. SHEAR REINFORCEMENT.

10.2.10.2.1. General. Shear reinforcement should form an angle of 90° to 45° with the mid-plane of the beam.

Clause 5.4.2.2(1)

Shear reinforcement (Fig. 10.15) may consist of a combination of

(a) links enclosing the longitudinal tensile reinforcement and the compression zone
(b) bent-up bars
(c) shear assemblies of cages, ladders, etc. which do not enclose the longitudinal

Clause 5.4.2.2(2)

DETAILING

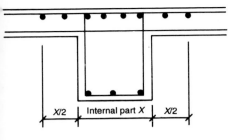

Fig. 10.14. Distribution of support reinforcement

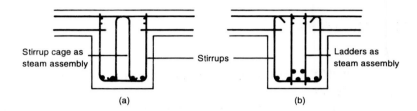

Fig. 10.15. Combination of links and shear assemblies: (a) stirrup cage as shear assembly; (b) ladders as shear assembly

reinforcement but are properly anchored in the compression and tension zones.

All shear reinforcement should be effectively anchored. Lap joints on the leg near the surface of the web are permitted only for high-bond bars. *Clause 5.4.2.2(3)*

At least 50% of the necessary shear reinforcement should be in the form of links. *Clause 5.4.2.2(4)*

10.2.10.2.2. Minimum area A_{sw}. $\rho_w = A_{sw}/sb_w \sin \alpha$ (Fig. 10.16), where ρ_w is the shear reinforcement ratio, A_{sw} is the area of shear reinforcement within length s, and α is the angle between the shear reinforcement and the longitudinal steel. Minimum values for ρ_w are given in Table 10.11. *Clause 5.4.2.2.(5)*

10.2.10.2.3. Maximum diameter. Diameter of reinforcement should not exceed 12 mm where plain round bars are used. *Clause 5.4.2.2(6)*

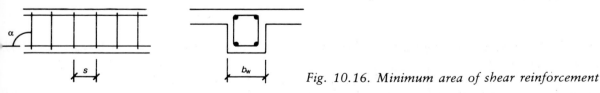

Fig. 10.16. Minimum area of shear reinforcement

Table 10.11. Minimum values for ρ_w

Concrete classes	Steel classes	
	S250	S460
C12/15 and C20/25	0·0014	0·0008
C25/30–C35/45	0·0021	0·0011
C40/50–C50/60	0·0026	0·0014

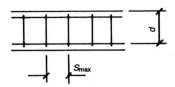

Fig. 10.17. Maximum longitudinal spacing of links and shear assemblies

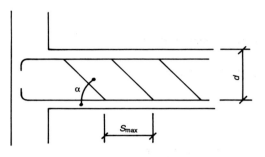

Fig. 10.18. Maximum longitudinal spacing of bent-up bars

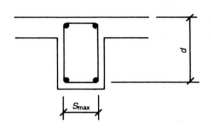

Fig. 10.19. Maximum transverse spacing of shear link legs

10.2.10.2.4. Maximum spacing S_{max}. See Fig. 10.17 for the maximum longitudinal spacing of links and shear assemblies.

$$V_{Sd} \leq \tfrac{1}{5} V_{Rd2}: S_{max} = 0\cdot8d \not> 300 \text{ mm}$$

$$\tfrac{1}{5} V_{Rd2} < V_{Sd} \leq \tfrac{2}{3} V_{Rd2}: S_{max} = 0\cdot6d \not> 300 \text{ mm}$$

$$V_{Sd} > \tfrac{2}{3} V_{Rd2}: S_{max} = 0\cdot3d \not> 200 \text{ mm}$$

where V_{Sd} is the design shear force and V_{Rd2} is the maximum shear force that can be carried by concrete.

See Fig. 10.18 for the maximum longitudinal spacing of bent-up bars.

$$S_{max} = 0\cdot6d\,(1 + \cot\alpha)$$

See Fig. 10.19 for the maximum transverse spacing of shear link legs.

$$V_{sd} \leq \tfrac{1}{5} V_{Rd2}: S_{max} = d \text{ or } 800 \text{ mm, whichever is smaller}$$

$$V_{sd} > \tfrac{1}{5} V_{Rd2}: \text{ as for longitudinal spacing}$$

Clause 5.4.2.2(7)

Clause 5.4.2.2(8)

Clause 5.4.2.2(9)

10.2.10.3. CURTAILMENT OF LONGITUDINAL REINFORCEMENT. Any curtailed reinforcement should be provided with an anchorage length $l_{b,net}$, but not less than d from the point where it is no longer needed. This should be determined taking into account the tension caused by the bending moment and that implied in the truss analogy used for shear design. This can be done by shifting the point of the theoretical cut-off based on the bending moment by a_1 (see below for definition) in the direction of decreasing moment. This procedure is also referred to as the 'shift rule'.

If the shear reinforcement is calculated according to the standard method

$$a_1 = z(1-\cot\alpha)/2 \not< 0$$

where α is the angle of the shear reinforcement to the longitudinal axis. If the shear reinforcement is calculated according to the variable strut method

$$a_1 = z(\cot\theta - \cot\alpha)/2 \not< 0$$

where θ is the angle of the concrete struts to the longitudinal axis. Normally z can be taken as $0 \cdot 9d$.

For reinforcement in the flange, placed outside the web, a_1 should be increased by the distance of the bar from the web.

10.2.10.4. ANCHORAGE AT SUPPORTS.

10.2.10.4.1. End support. When there is little or no fixity at an end support, at least a quarter of the span reinforcement should be carried through to the support. EC2 recommends that the bottom reinforcement should be anchored to resist force of $(V_{sd}a_1/d) + N_{sd}$, where V_{sd} is the shear force at the end, a_1 is as defined in section 10.2.10.3 for the shift rule and N_{sd} is the axial force, if any, in the member.

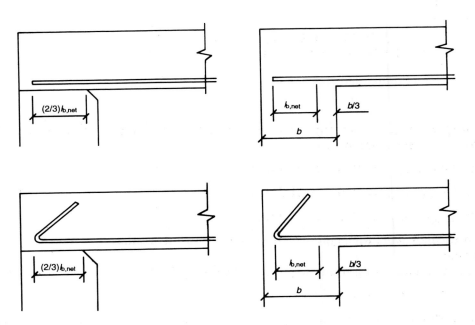

Fig. 10.20. Anchorage requirements

EC2 goes on to illustrate the anchorage requirement in *Fig. 5.12*, which arbitrarily reduces the anchorage requirement to $0 \cdot 67 l_{bnet}$ for direct supports. Clearly there is a presumption of adequate lateral pressure. It may be safer to use the formula in section 10.2.4.2 and arrive at the anchorage requirements. *Fig. 5.12* in the Code is reproduced here as Fig. 10.20, but it must be realized that l_{bnet} for curved bars is 70% of that for straight bars. The anchorages' length should be measured as in Fig. 10.20, and should be l_{bnet}.

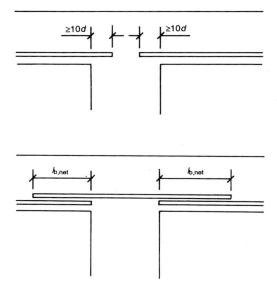

Fig. 10.21. Continuous reinforcement

10.2.10.4.2. Intermediate supports: general requirements. At intermediate supports, ≥ 25% of the mid-span bottom reinforcement should be carried to the support.

If no lacer bars are provided, bottom reinforcement should be anchored a minimum of 10ϕ beyond the face of the support. This does not mean that the support must be greater than 20ϕ wide, as the bars from each side of the support can be lapped. However, it is recommended that continuous reinforcement be provided to resist accidental forces (see Fig. 10.21).

Clause 5.4.2.1.4(1)

10.2.10.5. SKIN REINFORCEMENT. Skin reinforcement to control cracking should normally be provided in beams over 1·0 m in depth where the reinforcement is concentrated in a small portion of the depth. This reinforcement should be evenly distributed between the level of the tension steel and the neutral axis, and be located within the links.

Clause 5.4.2.4(2)
Clause 4.4.2.3(4)

10.2.10.6. SURFACE REINFORCEMENT. Surface reinforcement may be required to resist spalling of the cover, for example arising from fire or where bundled bars or bars greater than 32ϕ are used.

This reinforcement should consist of small-diameter high-bond bars or wire mesh placed in the tension zone outside the links.

The area of surface reinforcement parallel to the beam tension reinforcement should not be less than $0 \cdot 01 A_{ct,ext}$, where $A_{ct,ext}$ is the area of concrete in tension external to the links.

Clause 5.4.2.4

The longitudinal bars of the surface reinforcement may be taken into account as longitudinal bending reinforcement and the transverse bars as shear reinforcement, provided they meet the arrangement and anchorage requirements of these types of reinforcement (see Fig. 10.22).

Clause 5.4.2.4

10.2.11. Slabs

10.2.11.1. MINIMUM DIMENSIONS. Minimum overall depth for solid slabs = 50 mm (see Fig. 10.23).

Clause 5.4.3.1(1)

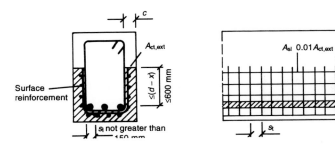

Fig. 10.22. Arrangement and anchorage requirements

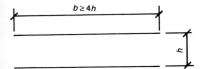

Fig. 10.23. Minimum overall depth for solid slabs

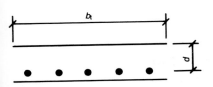

Fig. 10.24. Minimum area of longitudinal reinforcement

10.2.11.2. LONGITUDINAL REINFORCEMENT.

10.2.11.2.1. *Minimum area* $A_{st,min}$. *Clause 5.4.3.2.1(3)*

$$A_{st,min} \not< \frac{0 \cdot 6 b_t d}{f_{yk}} \not< 0 \cdot 0015 \, b_t d$$

where f_{yk} is the characteristic yield stress of reinforcement (see Fig. 10.24).

10.2.11.2.2. *Maximum area* $A_{st,max}$. *Clause 5.4.3.2.1(3)*

$$A_{st,max} \not> 0 \cdot 04 \, A_c$$

where A_c is the cross-sectional area of concrete.

10.2.11.2.3. *Maximum spacing* S_{max}. *Clause 5.4.3.2.1(4)*

$$S_{max} \not> 1 \cdot 5h \not> 350 \text{ mm}$$

See Fig. 10.25.

10.2.11.2.4. *Reinforcement near supports.* Span reinforcement: minimum 50% of the reinforcement in the span should be anchored at supports (see Fig. 10.26). *Clause 5.4.3.2.2(1)*

End supports with partial fixity, but simple support is assumed in design (see Fig. 10.27). *Clause 5.4.3.2.2(2)*

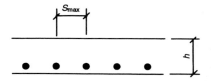

Fig. 10.25. *Maximum spacing of longitudinal reinforcement*

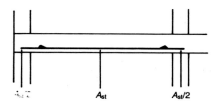

Fig. 10.26. *Span reinforcement*

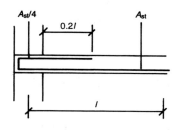

Fig. 10.27. *End supports with partial fixity*

10.2.11.2.5. Curtailment rules for slabs. These are similar to those for beams.

10.2.11.3. TRANSVERSE REINFORCEMENT.

10.2.11.3.1. Minimum area A_s. See Fig. 10.28.

Clause 5.4.3.2.1(2)

10.2.11.3.2. Maximum spacing.

$$S_{max} \not> 3h \not> 400 \text{ mm}$$

See Fig. 10.29.

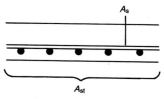

A_s not less than 20% A_{st} Fig. 10.28. *Minimum area of transverse reinforcement*

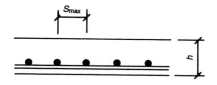

Fig. 10.29. *Maximum spacing of transverse reinforcement*

DETAILING

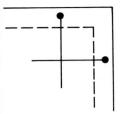

Fig. 10.30. *Corner reinforcement*

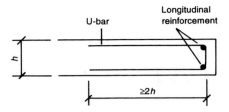

Fig. 10.31. *Reinforcement at free edges*

10.2.11.4. CORNER REINFORCEMENT. Suitable reinforcement is required where slab corners are restrained against lifting (see Fig. 10.30).
  U-bars in each direction extend $0 \cdot 2l$ into span.

Clause 5.4.3.2.3(1)

10.2.11.5. REINFORCEMENT AT FREE EDGES. See Fig. 10.31.

Clause 5.4.3.2.4(1)

10.2.11.6. SHEAR REINFORCEMENT.

10.2.11.6.1. Minimum slab depth. $h \not< 200$ mm where shear reinforcement is to be provided.

Clause 5.4.3.3(1)

10.2.11.6.2. General. The requirements given in section 10.2.10.2 for beams apply generally to slabs, with the following modifications.
  Form of shear reinforcement: shear reinforcement may consist entirely of bent-up bars or shear assemblies where

$$V_{Sd} \not> \tfrac{1}{3} V_{Rd2}$$

Maximum spacing for links: see Fig. 10.32

$$V_{Sd} \leq \tfrac{1}{5} V_{Rd2}: S_{max} = 0 \cdot 8d$$
$$\tfrac{1}{5} V_{Rd2} < V_{Sd} \leq \tfrac{2}{3} V_{Rd2}: S_{max} = 0 \cdot 6d$$
$$V_{Sd} > \tfrac{2}{3} V_{Rd2}: S_{max} = 0 \cdot 3d$$

Maximum spacing for bent-up bars: see Fig. 10.33.
  Shear reinforcement near supports for links: see Fig. 10.34.
  Shear reinforcement near supports for bent-up bars: see Fig. 10.35.
  Where a single line of bent-up bars is provided, their slope may be reduced to 30°.
  It may be assumed that one bent-up bar carries the shear force over a length of $2d$.

Clause 5.4.3.3(2)

Clause 5.4.3.3(3)

Clause 5.4.3.3(4)

Clause 5.4.3.3(5)

Clause 5.4.3.3(6)

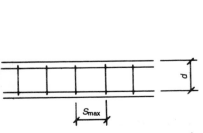

Fig. 10.32. *Maximum spacing for links*

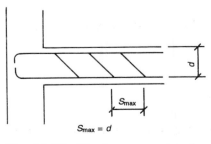

Fig. 10.33. *Maximum spacing for bent-up bars*

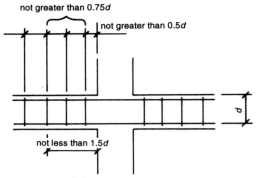

Fig. 10.34. Shear reinforcement near supports for links

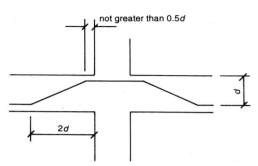

Fig. 10.35. Shear reinforcement near supports for bent-up bars

10.2.12. Columns

10.2.12.1. MINIMUM DIMENSIONS. See Fig. 10.36.

Clause 5.4.1
Clause 5.4.1.1

10.2.12.2. LONGITUDINAL REINFORCEMENT.

10.2.12.2.1. Minimum diameter. Minimum diameter is 12 mm.

Clause 5.4.1.2.1(1)

10.2.12.2.2. Minimum area $A_{s,min}$. See equation (5.13).

Clause 5.4.1.2.1(2)

$$A_{s,min} = \frac{0 \cdot 15 \, N_{sd}}{f_{yd}} \not< 0 \cdot 003 \, A_c$$

where N_{sd} is the design axial force, f_{yd} is the yield strength of reinforcement and $A_c = bh$.

10.2.12.2.3. Maximum area $A_{s,max}$.

Clause 5.4.1.2.1(3)

$$A_{s,max} = 0 \cdot 08 \, A_c$$

This maximum value also applies at laps. This is purely a practical consideration.

10.2.12.2.4. Minimum number of bars. See Fig. 10.37.

Clause 5.4.1.2.1(4)

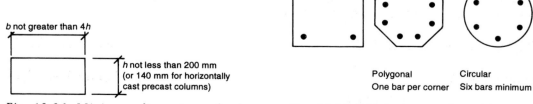

Fig. 10.36. Minimum dimensions of columns   Fig. 10.37. Minimum number of bars

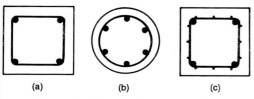

Fig. 10.38. *Minimum diameter of transverse reinforcement: (a) links; (b) helix; (c) mesh*

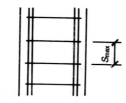

Fig. 10.39. *Maximum spacing of transverse reinforcement*

10.2.12.3. TRANSVERSE REINFORCEMENT.

10.2.12.3.1. General. All transverse reinforcement must be adequately anchored.

Every longitudinal bar (or group of bars) placed in a corner should be held by transverse reinforcement.

A maximum of five bars in or near each corner may be secured by any one set of transverse reinforcement. Although this is not stated in EC2, it will be advisable to limit the distance of the furthest bar from the corner to 150 mm.

Clause 5.4.1.2.2(2)
Clause 5.4.1.2.2(6)
Clause 5.4.1.2.2(7)

10.2.12.3.2. Minimum diameter. See Fig. 10.38.

$$\phi \not< \frac{\text{maximum } \phi \text{ main bars}}{4} \not< 6 \text{ mm}$$

if welded mesh fabric is used for transverse reinforcement ϕ of wires $\not< 5$ mm

Clause 5.4.1.2.2(1)

10.2.12.3.3. Spacing. General maximum spacing S_{max}: see Fig. 10.39.

Clause 5.4.1.2.2(3)

$$S_{max} \not> 12 \times \text{minimum } \phi \text{ of main bars}$$
$$\text{or } h$$
$$\text{or } 300 \text{ mm}$$

Figure 10.40 shows spacing at sections above and below slabs and beams.

Clause 5.4.1.2.2(4)

At lapped joints where maximum ϕ of main bars > 14 mm, reduced spacing of $0 \cdot 6 S_{max}$ should continue for the length of the lap.

Figure 10.41 shows spacing at changes in the direction of longitudinal bars.

Clause 5.4.1.2.2(5)

Spacing of transverse reinforcement should be calculated taking account of the forces generated by the change of direction.

10.2.13. Walls

10.2.13.1. MINIMUM DIMENSIONS. See Fig. 10.42. There is no EC2 requirement for this, but a practical minimum of 175 mm.

Clause 5.4.7.1(1)

10.2.13.2. VERTICAL REINFORCEMENT.

Clause 5.4.7.2(1)

10.2.13.2.1. Minimum area $A_{s_{v,min}}$.

Clause 5.4.7.2(2)

$$A_{s_{v,min}} \not< 0 \cdot 004 A_c$$

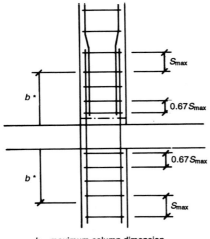

b = maximum column dimension
* With this height spacing of links to be reduced to 0.65 max.

Fig. 10.40. Spacing at sections above and below slabs and beams

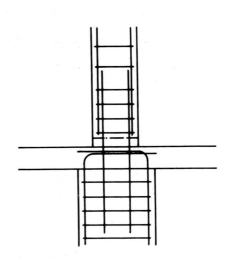

Fig. 10.41. Spacing at changes in the direction of longitudinal bars

10.2.13.2.2. Maximum area $A_{s_{v,max}}$. See Fig. 10.43.   *Clause 5.4.7.2(1)*
Clause 5.4.7.2(2)

$$A_{s_{v,max}} \not> 0.04 A_c$$

EC2 implies this to apply anywhere, including the laps.

10.2.13.2.3. Maximum spacing S_{max}. See Fig. 10.44.   *Clause 5.4.7.2(3)*

$$S_{max} \not> 2h \text{ or } 300 \text{ mm}$$

10.2.13.3. HORIZONTAL REINFORCEMENT.   *Clause 5.4.7.3(1)*

10.2.13.3.1. Placement. See Fig. 10.45. Horizontal reinforcement to be placed between vertical reinforcement and face of wall.

10.2.13.3.2. Minimum area $A_{s_{h,min}}$. See Fig. 10.46.   *Clause 5.4.7.3(1)*

$$A_{s_{h,min}} \not< \frac{A_{s_v}}{2}$$

10.2.13.3.3. Maximum spacing S_{max}. See Fig. 10.47.   *Clause 5.4.7.3(2)*

$$S_{max} = 300 \text{ mm}$$

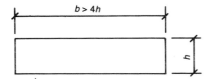

Fig. 10.42. Minimum dimensions of walls

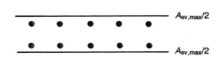

Fig. 10.43. Maximum area of vertical reinforcement

DETAILING

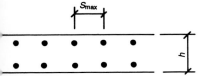

Fig. 10.44. Maximum spacing of vertical reinforcement

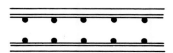

Fig. 10.45. Placement of horizontal reinforcement

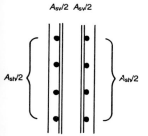

Fig. 10.46. Minimum area of horizontal reinforcement

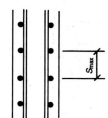

Fig. 10.47. Maximum spacing of horizontal reinforcement

10.2.13.3.4. *Minimum diameter*.   *Clause 5.4.7.3(3)*

$$\phi \text{ horizontal bars} \not< \frac{\phi \text{ vertical bars}}{4}$$

10.2.13.4. TRANSVERSE REINFORCEMENT. Where the area of vertical reinforcement exceeds $0 \cdot 02 A_c$, transverse reinforcement in the form of links should be provided in accordance with the requirements for columns.   *Clause 5.4.7.4*

10.2.14. Corbels

10.2.14.1. GENERAL. See Fig. 10.48. Where $0 \cdot 4 h_c \leq a_c \leq h_c$, a simple strut-and-tie model may be used.   *Clause 2.5.3.7.2*

10.2.14.2. TYING. Unless a length $l_{b,net}$ is available, the primary horizontal tie A_s should be anchored beyond the bearing area using U-bars or a welded crossbar (as shown in Fig. 10.48).   *Clause 5.4.4(1)*

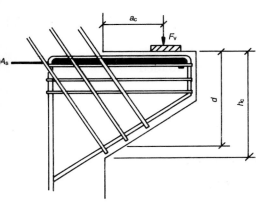

Fig. 10.48. General requirements for corbels

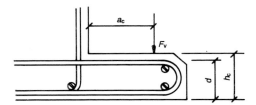

Fig. 10.49. General requirements for nibs

10.2.14.3. PROVISION OF LINKS. Where $h_c \geq 300$ mm and the area of the primary horizontal tie is such that $A_s \geq \boxed{0\cdot 4}\; A_c f_{cd}/f_{yd}$, where A_c is the sectional area of the concrete in the corbel at the column, close links with an area not less that $\boxed{0\cdot 4}\; A_s$ should be provided over the effective depth d.

Clause 5.4.4(2)

These links may be horizontal for a rectangular corbel and inclined for a tapering corbel.

10.2.15. Nibs

10.2.15.1. GENERAL. See Fig. 10.49. EC2 does not give any specific guidance for designing or detailing members that normally project from the faces of either beams or walls. Nibs less than 300 mm thick may be designed as short cantilevers. The line of action of load should be taken as the outer edge of the loaded area. The bending moment should be calculated on the line of the nearest vertical reinforcement in the member from which the nib projects, i.e. leg of a link in a beam or vertical reinforcement in a wall. Nibs should be sized so that shear reinforcement is not required. Provisions of Code *clause 4.3.2.2(9)* could be taken into account in assessing the shear resistance. The tension reinforcement should be adequately anchored by forming loops in the vertical or horizontal plane. Where nibs hang from the bottom of other members, sufficient tension reinforcement should be provided to transfer the loads from the nib to the member.

Clause 4.3.2.2(9)

10.2.16. Shear reinforcement in flat slabs

Guidance in EC2 on the description of shear reinforcement in flat slabs is scant compared to some national codes, e.g. the UK Code BS8110.

(a) Slabs where shear reinforcement is considered should be at least 200 mm thick.

(b) Shear should be checked at the first critical perimeter ($1\cdot 5d$ from the face of columns or loaded area). If the applied shear force on this perimeter exceeds the punching shear resistance of the slab without shear reinforcement, either the slab could be reinforced or the dimensions of the slab or the column should be amended to achieve the necessary resistance.

The percentage of tension reinforcement will be required in order to calculate the shear resistance. Only the reinforcement within a zone extending $1\cdot 5d$ or 800 mm (whichever is less) from the face of the loaded area should be taken into account.

The necessary shear reinforcement should be provided within the critical area, i.e. the area within the critical perimeter. Further critical perimeters should then be checked until a perimeter is reached where the resistance exceeds the applied shear force. EC2 does not stipulate the spacing of these further perimeters, but this may be taken as $0\cdot 75d$, as used in BS8110.

(c) EC2 does not specify how the shear reinforcement should be deployed within the critical area, except indirectly through *Fig. 5.17(a)*. Again

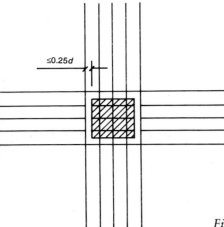

Fig. 10.50. Bent-up bars as shear reinforcement

following the British practice, which has proved satisfactory, the shear reinforcement calculated for each critical perimeter should be placed uniformly along two perimeters within $1 \cdot 5d$ from the perimeter where the reinforcement has been calculated. This would also comply with *Fig. 5.17(a)*.

The first set of link reinforcement should be placed $0 \cdot 5d$ from the face of the loaded end or column. The spacing of links should not exceed $0 \cdot 75d$.

(d) Shear reinforcement, when provided, should not be less than 60% of the values given in section 10.2.10.2.2 for beams.

(e) Bent-up bars may also be used as shear reinforcement. Where $V_{sd} \leq (1/3) V_{Rd2}$, the shear reinforcement may consist entirely of bent-up bars.

Only the bent-up bars over the loaded area or at a distance $0 \cdot 25d$ from the face of the loaded area should be taken into account. The distance from the face of the loaded area to the start of the crank of the bent-up bar, at the level of the tension reinforcement, should not exceed $0 \cdot 5d$ (see Fig. 10.50). The maximum spacing of bent-up bars is d.

10.2.17. Bundled bars

(a) Bars of the same diameter and characteristics may be bundled, and the rules for individual bars apply to the bundle.

(b) In design, equivalent diameter ϕ_n of the bundle should be used

$$\phi_n = \phi \sqrt{n_b} \geq 55$$

where n_b is the number of bars in a bundle, which is limited to four for vertical bars in compression and for bars in a lapped joint, and three for all other cases.

(c) Clear distances and concrete cover should be measured from the actual external contour of the bundle.

(d) The concrete cover should be greater than ϕ_n.

CHAPTER 11

Prestressed concrete

11.1. General

In EC2, prestressed concrete is treated as part of a much wider material group known as reinforced concrete, which covers normal reinforced concrete through partially prestressed concrete to fully prestressed concrete. The principles and methods given in EC2 are, in general, applicable to the full range of reinforced concrete construction. This chapter covers the aspects of design and detailing that are particular to prestressed concrete; those that apply equally to reinforced concrete are covered in other parts of this manual.

EC2: Part 1 covers the design of prestressed concrete members using only bonded internal tendons. Part 1.5 of EC2 will contain special rules applicable to unbonded and external tendons. At the time of writing, this had not yet been published, therefore this chapter considers only the design of members with internal bonded tendons.

11.2. Summary of main clauses

The main clauses relating to the design of prestressed concrete in EC2 are summarized in Table 11.1.

11.3. Durability

Steel reinforcement and prestressing tendons are protected against corrosion by complying with EC2 requirements on stress levels, crack widths and concrete cover. EC2 specifies a 'minimum' cover to which a construction tolerance of up to 10 mm must be added. For post-tensioned members, the minimum cover should not be less than the duct diameter. Although EC2 specifies requirements for the minimum cover perpendicular to the plane of curvature for curved bars with yield strengths up to 500 N/mm^2 (*Table 5.1*), it does not specify similar requirements for curved prestressing tendons. This is an unfortunate omission, and the designer is recommended to refer to other codes, e.g. BS8110, where specific requirements are stated for prevention of bursting of the cover perpendicular to the plane of curvature. The requirements of BS8110 are given in Table 11.2.

For pretensioned members, the minimum cover should be twice the tendon diameter, or three times the diameter when ribbed wires are used. The minimum cover is not related to the aggregate size, but this will probably not be a problem in practice when the construction tolerance is taken into account.

Clause 4.1.3.3(8)

Clause 4.1.3.3(5)
Clause 4.1.3.3(12)

Clause 4.1.3.3(11)

Table 11.1. Clauses for the design of prestressed concrete members

Requirement	Clause number
Durability	
Concrete cover	4.1.3.3
Design data	
Minimum concrete strength	4.2.3.5.2
Prestressing steel	
Mechanical properties	4.2.3.3
Minimum bending radii	4.2.3.3.6
Minimum number of wires	4.2.3.5.3
Jacking force	4.2.3.5.4
Ultimate limit state	
Bending and longitudinal force	4.3.1
Values of prestress	2.3.3.1, 2.5.4.2, 2.5.4.4
Shear	
General	4.3.2
Reduction in web width	4.3.2.2(8)
Shear capacity of concrete	4.3.2.3(1)
Maximum design shear stress	4.3.2.2(4), 4.3.2.3(3), 4.3.2.4.3(4), 4.3.2.4.4
Enhancement near supports	4.3.2.2(9), (10), (11)
Variable depth members	4.3.2.4.5
Inclined tendons	4.3.2.4.6
Torsion	4.3.3
Serviceability limit state	
Values of prestress	2.5.4.2, 2.5.4.3
Stress levels	4.4.1
Cracking	4.4.2
Shear reinforcement	4.4.2.3(5)
Deformation	4.4.3, Appendix 4
Torsion	4.4.2.3(5), 5.4.2.2, 5.4.2.3
Prestress losses	
Relaxation	4.2.3.4.1, 4.2.3.5.5(7), (9)
Elastic deformation	4.2.3.5.5(6)
Shrinkage	3.1.2.5.5, 4.2.3.5.5(9)
Creep	3.1.2.5.5, 4.2.3.5.5(9)
Draw-in	4.2.3.5.5(5)
Duct friction	4.2.3.5.5(8)
Anchorage zones	
Pretensioned members	4.2.3.5.6
Post-tensioned members	2.5.3.6.3, 2.5.3.7.4, 4.2.3.5.7, 5.4.6
Detailing	
Spacing of tendons/ducts	5.3.3
Anchorages and couplers	5.3.4
Minimum area of tendons	5.4.2.1.1
Tendon profile	5.4.2.1.3
Minimun shear reinforcement	5.4.2.2(5)
Spacing of shear reinforcement	5.4.2.2(7), (9)

In addition to these general requirements, the minimum cover depends on the condition of exposure to which the member will be subject (*Table 4.2*). In EC2, this requirement is independent of concrete strength, although the designer is referred to Table 3 of ENV 206 in order to select the appropriate concrete quality to use with the specified covers. Minimum covers for different conditions of exposure are given in Table 11.3.

Table 11.2. Minimum cover to curve ducts

Radius of curvature of duct	Duct internal diameter: mm															
	19	30	40	50	60	70	80	90	100	110	120	130	140	150	160	170
	Tendon force: kN															
	296	387	960	1337	1920	2640	3360	4320	5183	6019	7200	8640	9424	10 338	11 248	13 200
m	mm	mm	mm	mm	mm	mm	mm	mm	mm	mm	mm	mm	mm	mm	mm	mm
2	50	55	155	220	320	445								Radii not normally used		
4		50	70	100	145	205	265	350	420							
6			50	65	90	125	165	220	265	310	375	460				
8				55	75	95	115	150	185	220	270	330	360	395		
10				50	65	85	100	120	140	165	205	250	275	300	330	
12					60	75	90	110	125	145	165	200	215	240	260	315
14					55	70	85	100	115	130	150	170	185	200	215	260
16					55	65	80	95	110	125	140	160	175	190	205	225
18					50	65	75	90	105	115	135	150	165	180	190	215
20						60	70	85	100	110	125	145	155	170	180	205
22						55	70	80	95	105	120	140	150	160	175	195
24						55	65	80	90	100	115	130	145	155	165	185
26						50	65	75	85	100	110	125	135	150	160	180
28							60	75	85	95	105	120	130	145	155	170
30							60	70	80	90	105	120	130	140	150	165
32							55	70	80	90	100	115	125	135	145	160
34							55	65	75	85	100	110	120	130	140	155
36							55	65	75	85	95	100	115	125	140	150
38							50	60	70	80	90	105	115	125	135	150
40	50	50	50	50	50	50	50	60	70	80	90	100	110	120	130	145

Notes.
1. The tendon force shown is the maximum normally available for the given size of duct (taken as 80% of the characteristic strength of the tendon).
2. Where tendon profilers or spacers are provided in the ducts and these are of a type which will concentrate the radial force, the values given in the table will need to be increased.
3. The cover for a given combination of duct internal diameter and radius of curvature shown in the table, may be reduced in proportion to the square root of the tendon force when this is less than the value tabulated, subject to a minimum value of 50 mm.

Table 11.3. Cover to prestressing tendons and ducts: mm

Exposure class	Cover		Max. water/ cement ratio	Min. cement content: kg/m³
	Min.	nominal*		
1	25	30	0·60	300
2a	30	35	0·60	300
2b	35	40	0·55	300†
3	50	55	0·50	300†
4a	50	55	0·55	300
4b	50	55	0·50	300†
5a	35	40	0·55	300
5b	40	45	0·50	300
5c	50	55	0·45	300

* Includes a construction tolerance of 5 mm.
† Air entrainment required.

11.4. Design data

11.4.1. Concrete

The minimum concrete strength for use in pretensioned members is C30/37 and for post-tensioned construction is C25/30. The other design data to be assumed for

clause 4.2.3.5.2(1)

PRESTRESSED CONCRETE

Table 11.4. Concrete design data

Concrete class	C12/15	C16/20	C20/25	C25/30	C30/37	C35/45	C40/50	C45/55	C50/60	
f_{ck}	12	16	20	25	30	35	40	45	50	N/mm²
f_{ctm}	1·6	1·9	2·2	2·6	2·9	3·2	3·5	3·8	4·1	N/mm²
$f_{ctk0·05}$	1·1	1·3	1·5	1·8	2·0	2·2	2·5	2·7	2·9	N/mm²
$f_{ctk0·95}$	2·0	2·5	2·9	3·3	3·8	4·2	4·6	4·9	5·3	N/mm²
E_{cm}	26	27·5	29	30·5	32	33·5	35	36	37	kN/mm²

concrete are the same for prestressed members as for normal reinforced members, and are given in Table 11.4.

11.4.2. Prestressing steel

EC2 adopts a bilinear stress–strain diagram (*Fig. 4.6*) for prestressing steel with either a horizontal or sloping top leg, as shown in Fig. 11.1.

EC2 contains requirements (*Table 4.4*) on the minimum radii to which tendons of different composition can be bent: these are reproduced in Table 11.5.

The important issue of reliability is addressed by including simple requirements on the minimum number of bars, wires and tendons in the precompressed zone of isolated members, i.e. 'members in which no additional load-carrying capacity due to redistribution of internal forces and moments, transverse redistribution of loads or due to other measures (e.g. normal steel reinforcement) exists'. Seven-wire strands meet the reliability requirement, provided that the minimum wire diameter is greater than 4 mm.

clause 4.2.3.5.3

The maximum initial stress applied to the tendon, i.e. the jacking stress, is expressed as a percentage of the ultimate strength f_{pk} or the 0·1% proof stress $f_{p0·1k}$ of the tendon. In practice, $f_{p0·1k} \approx 0·85 \times f_{pk}$, and the requirement expressed in terms of the 0·1% proof stress will be the more critical. Similar requirements apply to the stress in the tendon after transfer. Values of jacking loads for various strand types are given in Table 11.6, where the requirement based on the 0·1% proof stress governs.

clause 4.2.3.5.4

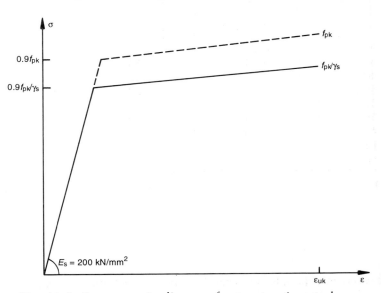

Fig. 11.1. *Stress–strain diagram for prestressing steel*

Table 11.5. Minimum radii for curved tendons

Type of tendon	Minimum bending radius
	Nominal diameter
Single wire or strand, deflected after tensioning	15
Single wire or strand, tensioned in smooth duct	20
Single wire or strand, tensioned in ribbed duct	40
Multi-wire or strand tendon	Preceding values multiplied by n_1/n_2

n_1 = total number of wires or strands in the tendon.
n_2 = number of wires or strands transferring the radial force of all wires or strands in the tendon to the deviator.

Table 11.6. Jacking stresses and initial prestress for different tendons: N/mm²

Strand	Dia.: mm	f_{pk}: N/mm²	$f_{p0 \cdot 1k}$: N/mm²	Jacking Stress	Initial Prestress
STD	15·2	1670	1420	1278	1207
	≤12·5	1770	1500	1350	1275
SUP	15·2	1770	1500	1350	1275
	≤12·9	1860	1580	1422	1343
DYF	18·0	1700	1450	1305	1233
	15·2	1820	1545	1391	1313
	12·7	1860	1580	1422	1343

f_{pk} = ultimate strength of tendon.
$f_{p0 \cdot 1k}$ = 0·1% proof stress of tendon.

11.5. Design of sections for flexure and axial load

11.5.1. Ultimate limit state

For the analysis of structures, EC2 adopts a partial safety factor for prestressing force γ_p of 1·0, while for the design of sections $\gamma_p = 1 \cdot 0$ only provided that both the following conditions are met:

clause 2.5.4.4

(a) not more than 25% of the total area of prestressed steel is located within the compression zone at the ultimate limit state
(b) the stress at the ultimate limit state in the prestressing steel closest to the tension face exceeds $f_{p0 \cdot 1k}/\gamma_m$.

If these conditions are not met, a value of $\gamma_p = 0 \cdot 9$ should be used in order to take account of variations across the section of the prestrain in the prestressing steel corresponding to a concrete stress $\sigma_c = 0$. These could be significant for sections near points of contraflexure where, because these sections are subjected to very low bending moments, the tendons may be spread across the section and experience a considerable variation in stress due to applied loads.

In practice, most prestressed members will meet both conditions (a) and (b) at their critical sections, so that γ_p can be taken as 1·0 in most cases.

PRESTRESSED CONCRETE

11.5.2. Serviceability limit state

EC2 distinguishes between three serviceability load combinations, i.e. rare, frequent and quasi-permanent. All are relevant to the design of prestressed concrete members, but their relative magnitude depends on the loads applied and the type of structure being designed.

EC2 also adopts an upper or lower characteristic value for the prestressing force (whichever is more critical) when crack widths and tendon stresses are being checked. The upper and lower characteristic values are taken as $1 \cdot 1$ and $0 \cdot 9$ times the mean value of the prestress respectively, provided the sum of the friction and long-term losses is less than 30%. If this condition is not met, EC2 does not offer any suggestions as to what values the designer should adopt. The designer is recommended to refer to section 4.2.6.2 of the draft CEB–FIP Model Code 1990.

The mean value of the prestressing force is used in the checking of concrete stresses.

Clause 2.5.4.3

Clause 2.5.4.2

11.5.2.1. LIMITATION ON STRESS.
To avoid the occurrence of longitudinal cracks in regions of high compression, EC2 limits the compressive stress under the rare load combination to $0 \cdot 6 f_{ck}$ in areas exposed to environments of exposure class 3 or 4 unless other measures, such as an increase in cover to reinforcement in the compressive zone or confinement by transverse reinforcement, are taken. In order to ensure that creep deformation is within the limits predicted by other parts of EC2, the concrete stress under the quasi-permanent load combination is limited to $0 \cdot 45 f_{ck}$. EC2 suggests that it may be necessary to check this condition at transfer, if creep at this stage is likely to affect significantly the functioning of the member. Otherwise, it is suggested that compressive stresses at transfer are limited to $0 \cdot 6 f_{ck}$.

To avoid stresses in the tendons under serviceability conditions that could lead to inelastic deformation of the tendons, EC2 limits the stress to $0 \cdot 75 f_{pk}$, after allowance for losses. This is unlikely to be a critical criterion in most cases.

EC2 draws the designer's attention to the fact that, in partially prestressed members, creep and shrinkage can lead to high stresses in both normal reinforcement and prestressing tendons, which could result in fatigue problems. This is a useful reminder to the designer.

Clause 4.4.1.1

Clause 4.4.1.1(3)

Clause 4.4.1.1(7)

Clause 4.4.1.2(2)

11.5.2.2. CRACKING.
EC2 specifies two criteria for the control of cracking: decompression and limiting crack widths to $0 \cdot 2$ mm (see Table 11.7). The decompression limit requires all parts of the tendon or duct to lie at least 25 mm within concrete in compression. All cracking checks for prestressed concrete are carried out under the frequent load combination using the upper or lower characteristic prestressing force.

Clause 4.4.2.1(7)

Table 11.7. Criteria for controlling cracking

Exposure class	Design crack widths, w_k, under the frequent load combination: mm	
	Post-tensioned	Pretensioned
1	0·2	0·2
2	0·2	
3	decompression or coating of the tendons and $w_k = 0 \cdot 2$	decompression
4		

221

In order to limit crack widths, EC2 tabulates bar size and spacing criteria (*Tables 4.11 and 4.12*) to ensure that widths are limited to 0·2 mm, as long as a minimum area of reinforcement or prestressing tendons is provided to control cracking due to the restraint of imposed deformations. Alternatively, formulae are provided to allow the design crack width to be calculated. The minimum area of reinforcement to control cracking arising from the restraint of imposed deformations can be reduced by taking account of the contribution of the prestressing tendons.

The approach adopted by EC2 for the design of partially prestressed concrete may be more rigorous than the approach adopted by some other codes, but it is more complex and difficult for the designer to apply. First, it should be noted that *Tables 4.11 and 4.12* apply only to *high-bond* bars. Prestressing strands have bond characteristics that are significantly less effective than high-bond bars, according to *clause 4.4.2.4(4)*. It is theoretically possible to use *Table 4.11* by determining an equivalent high-bond bar diameter from

Clause 4.4.2.4(4)

$$\phi' = \frac{k_{lps}}{k_{lhb}} \times \phi$$

where ϕ' is the equivalent high-bond bar diameter, ϕ is the prestressing strand diameter, k_{lhb} is the value for k_1 for high-bond bars (0·8) given in *clause 4.4.2.4(3)*, and k_{lps} is the value of k_1 for prestressing strand (2·0) given in *clause 4.4.2.4(4)*.

Clause 4.4.2.4(3)

Clause 4.4.2.4(4)

However, this will give equivalent diameters outside the range of *Table 4.11* for most normal strand diameters. Table 11.8 has been constructed by extending the approach used to derive *Table 4.11* so that it applies to prestressing strand, and this can be used in order to limit crack widths in prestressed concrete members.

Alternatively, the designer must use the crack width formulae. The procedure is as follows.

Clause 4.4.2.4

(a) Determine the moments for the frequent combination and the design value of the prestress force ($0·9P_{m,\infty}$).
(b) Calculate the cracking moment and hence determine σ_{sr}, the stress in the tension reinforcement calculated on the basis of a cracked section. EC2 does not explicitly state whether the lower (5%) fractile value or the mean

Table 11.8. Maximum tendon diameters for prestressing strands (crack width = 0·2 mm)

Steel stress: N/mm²	Maximum bar size: mm
80	40
90	32
100	25
115	20
130	16
150	12
160	10
180	8
210	6
230	5
260	4

The steel stress should be calculated for the frequent load combination on the basis of a cracked section regarding the prestress as an external force without allowing for the stress increase in the tendons due to loading.

value of the flexural tensile strength of concrete should be used for this purpose. As the aim is to limit the size of any cracks, it would seem reasonable to adopt the lower value of flexural tensile strength (Table 11.4). Once it is accepted that the section is cracked, the difference between the two values is unlikely to be significant as it affects only the degree of tension stiffening taken into account.

(c) Calculate the neutral axis depth x and the stress in the tension reinforcement under the frequent load combination σ_s for the cracked section.
(d) Calculate the mean strain ϵ_{sm}, using *equation 4.81*. β_1 should be taken as $0 \cdot 5$ for prestressing strand and β_2 should be taken as $0 \cdot 5$ for frequent loading.
(e) Calculate the average final crack spacing s_{rm} from *equation 4.82*. k_1 should be taken as $2 \cdot 0$ for prestressing strand and k_2 as $0 \cdot 5$ for bending. Note that s_{rm} is limited to $(h - x)$ in *clause 4.4.2.4(8)*, where h is the overall depth of the section. *Clause 4.4.2.4(8)*
(f) Calculate the design crack width w_k using *equation 4.80*.
(g) Compare w_k with the required value, and adjust the prestressing force and/or its eccentricity.
(h) Repeat steps (a)–(g) until the required value of w_k is obtained.

In prestressed concrete sections that also contain ordinary reinforcement, this can be used to control crack widths rather than relying on the prestress. In this case, *Tables 4.11 and 4.12* will be applicable.

11.5.2.3. DESIGN EQUATIONS.
In order to assist the designer, the stress and crack width criteria discussed in the previous sections can be expressed mathematically as a series of inequalities. The following nomenclature is used

A	= area of concrete section
e	= eccentricity of tendon
f_{ck}	= 28-day cylinder strength
f_{ctm}	= mean tensile strength of concrete
M_{DL}	= moment acting at transfer
M_F	= moment under frequent loads
M_{QP}	= moment from quasi-permanent loads
M_R	= moment under rare loading
P_{m0}	= prestress force at transfer $(t = 0)$
$P_{m\infty}$	= final prestress force $(t = \infty)$
Z_b	= section modulus of bottom fibre
$Z_{b(25)}$	= section modulus 25 mm below the tendon
Z_t	= section modulus of top fibre

Sagging moments are positive and e is taken as positive when it is below the neutral axis.

11.5.2.3.1. Lower chord, compression at transfer.

$$\frac{A}{P_{m0}} \geq \frac{\left(1 + \dfrac{Ae}{Z_b}\right)}{0 \cdot 45 (\text{or } 0 \cdot 6) f_{ck} + \dfrac{M_{DL}}{Z_b}} \quad \text{for} \quad P_{m0}\left(\frac{1}{A} - \frac{e}{Z_t}\right) + \frac{M_{DL}}{Z_t} \geq -f_{ctm}$$

Otherwise the calculation should be based on a cracked section.

11.5.2.3.2. Upper chord, tension at transfer. Check that the crack width does not exceed 0·2 mm. In this calculation γ_p should be taken as 1·1 in order to maximize the tensile stress, i.e. the prestressing force should be taken as $P = 1 \cdot 1 P_{m0}$.

11.5.2.3.3. Lower chord, tension at SLS. Check decompression:

$$\frac{A}{P_{m\infty}} \leq \frac{0 \cdot 9 \left(1 + \dfrac{Ae}{Z_{b(25)}}\right)}{\dfrac{M_F}{Z_{b(25)}}} \quad \text{for } 0 \cdot 9 P_{m\infty}\left(\frac{1}{A} + \frac{e}{Z_b}\right) - \frac{M_R}{Z_b} \geq -f_{ctm}$$

Otherwise the designer should use a cracked section and adjust P and e until the neutral axis is 25 mm below the tendon.

Check crack width: the crack width should be determined with $\gamma_p = 0 \cdot 9$ in order to maximize tension, i.e. $P = 0 \cdot 9 P_{m\infty}$. The bending moment should be M_F.

Check tendon stress: calculate the tendon stress using a cracked section. The bending moment should be M_R. The prestressing force $P = \gamma_p P_{m\infty}$. Both $\gamma_p = 0 \cdot 9$ and 1·1 should be considered, although the higher value is more likely to be critical.

11.5.2.3.4. Upper chord, compression at SLS. Check that:

$$\frac{A}{P_{m\infty}} \geq \frac{\left(1 + \dfrac{Ae}{Z_t}\right)}{0 \cdot 6 f_{ck} - \dfrac{M_R}{Z_t}} \quad \text{for } P_{m\infty}\left(\frac{1}{A} + \frac{e}{Z_b}\right) - \frac{M_R}{Z_b} \geq -f_{ctm}$$

$$\frac{A}{P_{m\infty}} \geq \frac{\left(1 - \dfrac{Ae}{Z_t}\right)}{0 \cdot 45 f_{ck} - \dfrac{M_{QP}}{Z_t}} \quad \text{for } P_{m\infty}\left(\frac{1}{A} + \frac{e}{Z_b}\right) - \frac{M_R}{Z_b} \geq -f_{ctm}$$

Otherwise, the calculation must be carried out on a cracked section.

11.5.2.4. DEFORMATION. The span/depth ratios given in *Table 4.14* are stated to apply only to reinforced concrete. The designer is referred to the guidance given in *Appendix 4* on the calculation of deformations using elastic methods, when it is necessary to determine the deflections of prestressed concrete members.

Appendix 4

11.6. Design of sections for shear and torsion

11.6.1. Shear

11.6.1.1. ULTIMATE LIMIT STATE. In the design of sections for shear, EC2 follows the same procedure as for reinforced concrete, treating the prestressing force as an applied axial compression. A partial safety factor, $\gamma_p = 1 \cdot 0$, should be used unless the conditions given in *clause 2.5.4.4.3(3)* are not satisfied, in which case γ_p should be reduced to 0·9. If the area of prestressing tendons has been increased to satisfy the serviceability limit state for flexure, or the ultimate moment

Clause 4.3.2

Clause 2.5.4.4.3(3)

coincident with the design shear force is significantly less than the ultimate design moment, the stress at the ultimate limit state in the tendons closest to the tensile fibre may not exceed $f_{p0.1k}/\gamma_m$ and γ_p must be taken as 0.9.

One of the factors governing the shear capacity of the concrete is the area of longitudinal reinforcement in the tensile zone. EC2 does not state whether this should be the total area of both the normal reinforcement and the prestressing tendons or whether the areas should be adjusted to reflect the different design strengths. It is suggested that the sum of the areas be used, as this term is included to account for the change in level of the neutral axis, which is governed more by reinforcement area than by its strength.

As described in Chapter 6, the designer is allowed to choose between two methods for determining the shear capacity of a section with shear reinforcement. Both adopt a truss model for determining the capacity of the shear reinforcement. In the standard method, the concrete compression struts are assumed to act at 45° to the horizontal, and the shear capacity of the concrete section is added to the capacity of the shear reinforcement to determine the total capacity of the section.   *Clause 4.3.2.4.3*

In the variable strut inclination method, the designer is free, within certain limits, to choose the angle of the concrete compression strut. The area of shear reinforcement required to maintain equilibrium can then be calculated. In this method the shear capacity of the concrete is not added to the shear capacity of the truss, and consequently no benefit is gained from prestressing in the horizontal direction. However, this method can still lead to a more economical design of shear reinforcement in some situations, as demonstrated in Fig. 11.2 and Table 11.9.   *Clause 4.3.2.4.4*

The designer's attention is specifically drawn to the fact that the tensile force in the longitudinal reinforcement is increased by the presence of shear above that required for bending. Provision for this effect is generally made by shifting the bending moment diagram so that the moment at a given section is always increased. This is discussed in Chapter 10. Although the variable strut inclination method can lead to lower areas of shear reinforcement, there will be a corresponding increase in the longitudinal reinforcement requirement.

The maximum shear force that can be carried by a section is limited to that which will cause crushing of the concrete. If the web contains grouted ducts with a diameter greater than one-eighth of the web width, the effective web width should be reduced by 50% of the sum of the duct diameters at the most unfavourable level.   *Clause 4.3.2.2(8)*

The maximum shear force is reduced when the section is subjected to an axial compressive force. Although this is not specifically stated in EC2, it is suggested that this reduction should also apply to any section subjected to a prestressing force. In this situation the prestressing force is an unfavourable effect, and a partial safety factor of $\gamma_p = 1.2$ should be used.   *Clause 4.3.2.2(4)*

EC2 allows for the increased resistance of prestressed concrete sections close to direct supports due to the direct transmission of loads in the same way as for reinforced concrete. However, no such enhancement should be considered when checking the design shear force against the maximum shear capacity of the section V_{Rd2} or $V_{Rd2.red}$.   *Clause 4.3.2.2(9), (10), (11)*

Clause 4.3.2.2(5)

The effect on the design shear force (positive or negative) of inclined tendons and compression and tension zones must be included when calculating the design shear force. However, a reduction in the design shear force due to inclined compression or tension zones can be combined with a reduction due to inclined prestressing tendons only if a detailed verification can be given. The designer should select the appropriate partial safety factor, depending on whether the effect of the load is favourable or unfavourable. For prestressing tendons the appropriate values are as follows.   *Clause 4.3.2.4.5*
Clause 4.3.2.4.6

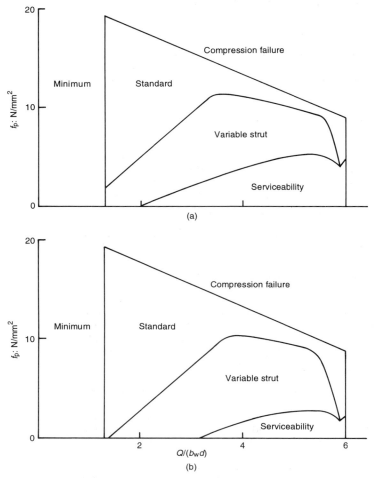

Fig. 11.2. Comparison of variable-strut and standard methods for determining shear capacity (concrete grade 40/50): (a) longitudinal reinforcement ratio $\rho = 0.0015$; (b) $\rho = 0.02$

When the stress in the tendons does not exceed $f_{p0.1k}$, $\gamma_p = 0.9$ for favourable effects and 1.2 for unfavourable effects.

When the stress in the tendons exceeds $f_{p0.1k}$, EC2 specifies that the prestressing force V_{pd} should be calculated assuming a stress of $f_{p0.1k}/\gamma_s$, which presumably applies for favourable effects. It does not specify the value to be used when the effect is unfavourable: it is suggested that a value of $\gamma_p = 1.2$ be applied to the calculated force.

11.6.1.2. SERVICEABILITY. Cracking under service loads due to shear is likely to occur when the ultimate design shear force is greater than three times the shear capacity of the concrete section without shear reinforcement. In order to limit crack widths, EC2 relates the excess stress in the shear reinforcement at the ultimate limit state to the maximum stirrup spacing (*Table 4.13*). This criterion is likely to be critical in members with a high applied ultimate shear stress and a low mean value of prestress, as illustrated in Fig. 11.2 and Table 11.9.

Clause 4.4.2.3(5)

Table 11.9. Comparison of 'variable strut' and standard methods for determining shear capacity — concrete grade 40/50

$Q/b_w d$: N/mm²	f_p: N/mm²	$A_{sw}/b_w s$		
		Variable strut		Standard
0	0	0·0015		0·0015
	5	0·0015		0·0015
	10	0·0015		0·0015
	15	0·0015		0·0015
	20	0·0015		0·0015
1	0	0·0019 to 0·0041		0·0015
	5	0·0019 to 0·0041		0·0015
	10	0·0019 to 0·0041		0·0015
	15	0·0019 to 0·0041		0·0015
	20	0·0028		0·0015
2	0	0·0037 to 0·0083		0·0033 to 0·0041
	5	0·0037 to 0·0083		0·0015 to 0·0022
	10	0·0037 to 0·0083		0·0015
	15	0·0037 to 0·0083		0·0015
	20	*		*
3	0	0·0056 to 0·0124		0·0061 to (0·0075)
	5	0·0056 to 0·0124		0·0042 to 0·0050
	10	0·0056 to 0·0124		0·0023 to 0·0031
	15	0·0056 to 0·0124		0·0015
	20	*		*
4	0	(0·0081) to 0·0166		0·0088 to (0·0128)
	5	0·0074 to 0·0166		0·0070 to 0·0078
	10	0·0074 to 0·0166		0·0051 to 0·0059
	15	*		*
	20	*		*
5	0	(0·0133) to (0·0181)		(0·0133) to (0·0181)
	5	0·0093 to 0·0167		0·0097 to 0·0106
	10	0·0093 to 0·0167		0·0079 to 0·0087
	15	*		*
	20	*		*
6	0	(0·0186) to (0·0234)		(0·0186) to (0·0234)
	5	0·0167		0·0125 to 0·0133
	10	*		*
	15	*		*
	20	*		*

* Shear capacity exceeded.
Areas governed by crack control criteria are shown in parentheses ().

11.6.2. Torsion

Torsional resistance must be calculated when the equilibrium of the structure depends on its torsional resistance. The design should be carried out at both the serviceability and ultimate limit states; the necessary design rules are given in EC2. The reinforcement is designed using the variable strut inclination method, which must also be used when designing for any coexistent shear forces.

Clause 4.3.3.1(1)

Clause 4.4.2.3(5)
Clause 4.3.3.1(6)
Clause 4.3.3.2.2

When torsion arises from consideration of compatibility only, EC2 requires the member to be designed to avoid excessive cracking. In practice this means that stirrups and longitudinal reinforcement should satisfy specified detailing rules.

Clause 5.4.2.2
Clause 5.4.2.3

11.7. Prestress losses

11.7.1. General
The designer's attention is drawn to the need to allow for losses of prestress when calculating the design forces in tendons at the various stages considered in the design. The causes of these losses are listed in EC2. Because of the uncertainty in estimating such losses, it is suggested that experimental evidence should be used where it is available. In the absence of such data, EC2 suggests values for the various parameters that can be used for design.

11.7.2. Friction in jack and anchorages
Loss of prestressing force due to friction in the jack and anchorages is not mentioned in EC2. It is not generally a problem for the designer, because the design is normally based on the tendon force on the concrete member side of the jack. However, it does need to be allowed for in calibration of the actual jack used for stressing.

11.7.3. Duct friction
Clause 4.2.3.5.5(8)

EC2 specifies the standard formula for calculating the force lost in overcoming duct friction when the tendon is stressed. This separates the loss into two parts

(a) that caused by the tendon and duct following the specified profile
(b) that caused by slight variations in the actual line of the duct, which may cause additional points of contact between the tendon and the duct wall.

The losses are calculated in terms of the coefficient of friction μ and an unintentional angular displacement per metre length k. Values of μ for a range of tendons which fill about 50% of the duct are given in EC2, which also suggests that k should be taken between 0·005 and 0·01 rad/m. Rather than using these values, it is generally better to refer to literature and test results produced by the manufacturers of prestressing components, and EC2 allows the designer to adopt this approach.

11.7.4. Elastic deformation
Clause 4.2.3.5.5(6)

The method specified for the calculation of losses due to the immediate elastic deformation of the concrete when the prestressing force is applied follows normal procedures.

11.7.5. Anchorage draw-in or slip
Clause 4.2.3.5.5(5)

EC2 mentions this cause of loss of prestressing force, but does not specify the values to be used. Appropriate values can be obtained from the anchorage manufacturers and should be checked on site, particularly if the member is short, when the loss due to this cause can be critical.

11.7.6. Time-dependent losses
Time-dependent losses are those due to relaxation of the prestressing tendons and shrinkage and creep of the concrete. EC2 considers the loss due to each effect as if that effect were acting alone, and then calculates the combined effect using an interaction formula. Because these losses are interdependent, it is necessary to use an iterative procedure. Only a few iterations are generally required in order to obtain a reasonable estimate of the total time-dependent loss.

Clause 4.2.3.5.5(9)

11.7.6.1. RELAXATION OF STEEL.
The starting point is the relaxation loss at 1000 h, which is best obtained from the test certificates for the tendons, or

Clause 4.2.3.4.1

Table 11.10. Relaxation losses as a percentage of initial stress (σ_{po}) at 20°C

Tendon type	Initial stress: σ_{po}	Loss after									
		1h	5h	20h	100h	200h	500h	1000h	long-term		
Class 1 (wires)	$0.6f_{pk}$	0.7	1.1	1.6	2.5	2.9	3.8	4.5	9.0	to	13.5
	$0.7f_{pk}$	1.2	2.0	2.8	4.4	5.2	6.8	8.0	16.0	to	24.0
	$0.8f_{pk}$	1.8	3.0	4.2	6.6	7.8	10.2	12.0	24.0	to	36.0
Class 2 (strands)	$0.6f_{pk}$	0.2	0.3	0.4	0.6	0.7	0.9	1.0	2.0	to	3.0
	$0.7f_{pk}$	0.4	0.6	0.8	1.4	1.6	2.1	2.5	5.0	to	7.5
	$0.8f_{pk}$	0.7	1.1	1.6	2.5	2.9	3.8	4.5	9.0	to	13.5
Class 3 (bars)	$0.6f_{pk}$	0.2	0.4	0.5	0.8	1.0	1.3	1.5	3.0	to	4.5
	$0.7f_{pk}$	0.6	1.0	1.4	2.2	2.6	3.4	4.0	8.0	to	12.0
	$0.8f_{pk}$	1.1	1.8	2.5	3.9	4.6	6.0	7.0	14.0	to	21.0

from the appropriate material standard. EC2 suggests values for this parameter in the absence of any other source of data. The 1000 h relaxation loss is then multiplied by a factor of 3 to give the long-term loss. A value of 3 is high for low relaxation strand, and will result in an overestimate of the relaxation loss. A value of 2 is considered to be more appropriate for low relaxation strand. Relaxation losses are given in Table 11.10.

11.7.6.2. SHRINKAGE. EC2 suggests typical values within the main text for the final shrinkage strain of concrete (*Table 3.4*), with more detailed information in *Appendix 1*.

Appendix 1

11.7.6.3. CREEP. EC2 suggests typical values within the main text for the final specific creep strain (or creep coefficient) for concrete (*Table 3.3*), with more detailed information in *Appendix 1*.

Appendix 1

11.8. Anchorage zones

11.8.1. Pretensioned members

Clause 4.2.3.5.6

Guidance is provided in EC2 for estimating the transmission lengths in pretensioned members. A distinction is drawn between the transmission length over which the prestressing force is fully transmitted to the concrete, the dispersion length over which the concrete stresses gradually disperse to a distribution compatible with plane sections remaining plane, and the anchorage length over which the tendon force at the ultimate limit state is fully transmitted to the concrete. The designer will find that this distinction helps to clarify the distribution of forces within the anchorage zone. Transmission lengths are given in Table 11.11.

Table 11.11. Transmission lengths: number of diameters

Tendon type	Concrete grade					
	25/30	30/37	35/45	40/50	45/55	50/60
Strands and smooth or indented wires	75	70	65	60	55	50
Ribbed wires	55	50	45	40	35	30

11.8.2. Post-tensioned members

Clause 4.2.3.5.7

The design of anchorage zones, or end blocks, in post-tensioned members is a very important area of design, but one that is rarely well-treated in codes of practice. EC2 is no exception. EC2 draws attention to this important area and outlines the principles of the design.

EC2 in its present draft is confusing: *clause 4.2.3.5.7* instructs the designer to check bearing stresses under the anchorage in accordance with *clause 5.4.8*, which specifically excludes prestressing anchorages. The designer's attention is drawn to the need to consider bearing stresses behind the anchorage, overall equilibrium and transverse tensile forces, i.e. bursting and spalling forces. The design is carried out at the ultimate limit state and is based on the characteristic strength of the prestressing tendon, not on γ_p times the prestressing force. Crack control is not mentioned, presumably being deemed to be covered in the general detailing requirements. It is suggested that overall equilibrium be checked using a strut-and-tie method, and guidance is given on this and on the angle of dispersion of the prestressing force into the concrete section. This is of benefit to the designer. However, no guidance is given on how to design against spalling and bursting forces. Some guidance is given on the detailing of anchorage zones, but the information given could be more precise.

Clause 4.2.3.5.7
Clause 5.4.8

Clause 2.5.3.6.3

Clause 2.5.3.7.4

Clause 5.4.6

The designer's attention is drawn to the importance of paying special attention to anchorage zones that have a cross-section of different shape from that of the general cross-section of the member.

For more information on the design and detailing of anchorage zones in prestressed concrete, the designer is referred to the excellent documents produced by CIRIA and the Institution of Structural Engineers.

11.9. Detailing

Section 5

The detailing provisions given in *section 5* apply to both prestressed and normally reinforced concrete. In this section, only provisions that specifically relate to prestressed concrete are discussed. A more general commentary on the detailing requirements of EC2 is given in Chapter 10.

11.9.1. Spacing of tendons and ducts

Clause 5.3.3

EC2 specifies the minimum allowable clear spacing between adjacent tendons or ducts, which is generally the greater of the tendon or duct diameter and the aggregate size plus 5 mm. Absolute minimum clear spacings are also specified, but these are unlikely to be critical when account is taken of the need to be able to compact the concrete around the tendons. For post-tensioned members, EC2 bases its requirements on the outside dimensions of the duct rather than on the internal dimensions, the latter being more normal British practice. The requirements are summarized in Table 11.12.

EC2 does not specify any additional requirements for the minimum clear spacing

Table 11.12. Minimum clear spacing requirements for prestressing tendons and ducts

Pretensioned	Vertically	$\geq d_g$	$\geq \phi$	≥ 10 mm
	Horizontally	$\geq d_g + 5$ mm	$\geq \phi$	≥ 20 mm
Post-tensioned	Vertically		$\geq \phi$	≥ 50 mm
	Horizontally		$\geq \phi$	≥ 40 mm

ϕ = diameter of tendon or duct, as appropriate.
d_g = aggregate size.

Table 11.13. Minimum distance between centre-lines of ducts in plane of curvature

Radius of curvature of duct	Duct internal diameter: mm															
	19	30	40	50	60	70	80	90	100	110	120	130	140	150	160	170
	Tendon force: kN															
	296	387	960	1337	1920	2640	3360	4320	5183	6019	7200	8640	9424	10 336	11 248	13 200
m	mm	mm	mm	mm	mm	mm	mm	mm	mm	mm	mm	mm	mm	mm	mm	mm
2	110	140	350	485	700	960								Radii not normally used		
4	55	70	175	245	350	480	610	785	940							
6	38	60	120	165	235	320	410	525	630	730	870	1045				
8			90	125	175	240	305	395	470	545	655	785	855	940		
10			80	100	140	195	245	315	375	440	525	630	685	750	815	
12						160	205	265	315	365	435	525	570	625	680	800
14						140	175	225	270	315	375	450	490	535	585	785
16							160	195	235	275	330	395	430	470	510	600
18								180	210	245	290	350	380	420	455	535
20									200	220	265	315	345	375	410	480
22											240	285	310	340	370	435
24												265	285	315	340	400
26												260	280	300	320	370
28																345
30																340
32																
34																
36																
38																
40	38	60	80	100	120	140	160	180	200	220	240	260	280	300	320	340

Notes.
1. The tendon force shown is the maximum normally available for the given size of duct (taken as 80% of the characteristic strength of the tendon).
2. Value less than 2 × duct internal diameter are not included.
3. Where tendon profilers or spacers are provided in the ducts and these are of a type which will concentrate the radial force, the values given in the table will need to be increased. If necessary reinforcement should be provided between ducts.
4. The distance for a given combination of duct internal diameter and radius of curvature shown in the table may be reduced in proportion to the tendon force when this is less than the value tabulated, subject to meeting the minimum requirements of Table 11.12.

in the plane of curvature between curved tendons. This is a serious omission, as very high local forces can be developed. The designer is recommended to use the requirements given in other codes, such as BS8110, which are reproduced in Table 11.13.

11.9.2. Anchorages and couplers

Clause 5.3.4

EC2 requires that not more than 50% of the tendons be coupled at any one cross-section, which is contrary to the generally accepted UK practice of coupling all the tendons at one cross-section. The reason for the EC2 requirement is that when a coupled tendon is stressed the force between the previously stressed concrete and the anchorage decreases, and the local deformation of this concrete is reduced. Increased compressive forces are induced adjacent to the anchorage, and balancing tensile forces between adjacent anchorages. Uncoupled tendons across the joint reduce this effect, as well as assisting in carrying the induced tensile forces. Alternatively, it is suggested that such forces could be carried by properly detailed normal reinforcement.

11.9.3. Minimum area of tendons

Clause 5.4.2.1.1

The minimum area of tendons is governed by the same requirement as for a normally reinforced concrete member, with f_{yk} replaced by f_{pk}. In all practical cases, this means that EC2 will require a minimum area of longitudinal reinforcement of 0·15%, which could be made up of normal reinforcement or prestressing tendons with no adjustment for their different characteristic strengths.

11.9.4. Tendon profiles

Clause 5.4.2.1.3

The design for shear in EC2 leads to a requirement for longitudinal reinforcement in addition to that required for bending. This reinforcement is generally allowed for by a horizontal displacement of the bending moment envelope (the 'shift' rule). When the flexural design of a prestressed concrete member is governed by the ultimate limit state, which could be the case when the design criterion is a limiting crack width of 0·2 mm, the longitudinal reinforcement will need to be increased or the tendon profile adjusted in order to meet the shear requirements.

11.9.5. Shear reinforcement

11.9.5.1. MINIMUM AREA. The minimum area of shear reinforcement depends on its yield strength and the compressive strength of the concrete. The requirements are summarized in Table 11.14. Concrete grades 12/15 and 20/25 are not allowed for prestressed concrete members.

Clause 5.4.2.2(5)

Minimum shear reinforcement is not required in slabs that have adequate provision for the transverse distribution of loads or in members of minor importance, e.g. a lintel with a span of less than 2 m.

11.9.5.2. MAXIMUM SPACING. The maximum longitudinal and transverse spacing of shear reinforcement depends on the magnitude of the applied shear force relative to the maximum design shear force V_{Rd2} that can be carried without crushing the notional concrete compressive struts. As the applied shear force increases, the maximum allowable spacing decreases. Absolute maximum spacings are also specified.

Clause 5.4.2.2(7)
Clause 5.4.2.2(9)

These criteria will encourage the adoption of smaller diameter shear links at closer centres. This will be beneficial in controlling any shear cracking that occurs, and may in any case be necessary in order to satisfy the crack control requirements.

Clause 4.4.2.3(5)

Table 11.14. Minimum shear reinforcement ratios ($A_{sw}/b_w s$)

Concrete grade	Steel grade		
	S220	S400	S500
12/15, 20/25	0·0016	0·0009	0·0007
25/30 to 35/45	0·0024	0·0013	0·0011
40/50 to 50/60	0·0030	0·0016	0·0013

APPENDIX

Appendix: Design of sway frames in EC2

A.1. Development of a simplified method

EC2 provides a simple method of designing isolated, braced columns for the effects of deflections. This is given in *clause 4.3.5.6.3*. No method is given for the simple design of sway frames but *clause A.3.5(2)* in *Appendix 3* states that: 'The simplified methods defined in 4.3.5 may be used instead of a refined analysis, provided that the safety level required is ensured'.

Clause 4.3.5.6.3
Clause A.3.5(2)
Appendix 3

A design method for sway frames using the model column approach is developed below. The basic approach is to develop the method given for sway frames in BS8110 to provide a more rigorously justifiable (and more economical) method.

The basic approach of the model column method is to define a deformed shape for the column under ultimate conditions, notably the situation where the strain in the concrete reaches its ultimate value (usually 0·0035). This poses no major problem for single columns, but is much more complex for sway frames.

Consider the structure shown in Fig. A.1. The fundamental point to note is that, at any stage, the sidesway deflection of all columns must be the same. It seems reasonable to postulate that failure occurs when the first column reaches its ultimate deformation.

The deformed shape of a column under the action of sidesway is of the form shown in Fig. A.2.

Where the column is rigidly fixed at top and bottom, the deflections corresponding to the ultimate condition can be calculated using the same equations as for isolated braced columns (Fig. A.3).

In fact, the ends will not be rigidly restrained and it is of interest to consider the influence of this. Consider sidesway of a column fixed at one end and pinned at the other (Fig. A.4). In this case, the effective length will be found to be twice the actual length while in the first case the actual and effective lengths are equal. Hence, it is convenient to use the same principles as for isolated columns, and take it that

$$a = \frac{1}{10} \frac{1}{r} l_o^2$$

For the sway frame, failure occurs when the first column reaches its ultimate deformation. The failure deformation for the structure is thus the minimum ultimate deflection for any of the columns. In Fig. A.1, assuming all the columns have the same section, this will be the ultimate deflection calculated for the shortest column (column 3). All the columns are subjected to this deflection. On the face of it, the ith column can now be designed for

$$M_i = N_i (e_{\min} + e_{o.i} + e_{a.i})$$
$$N = N_i$$

Unfortunately, there is a snag! Consider the moment–deflection characteristics of columns.

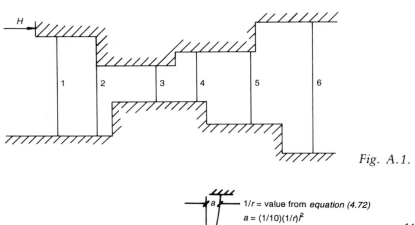

Fig. A.1.

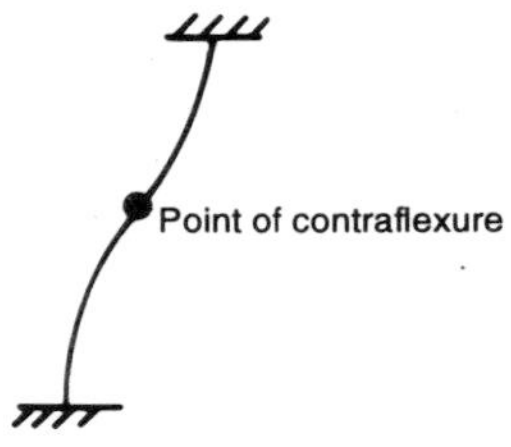

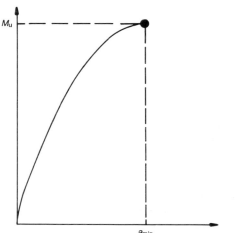

Fig. A.2. Deformed shape of a column under the action of sidesway

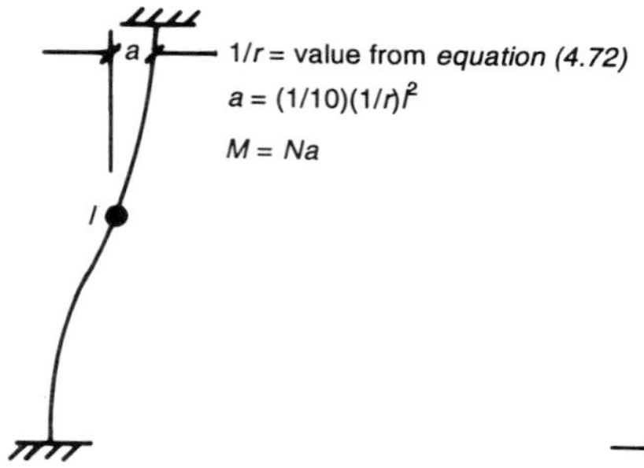

Fig. A.3. Column rigidly fixed at top and bottom

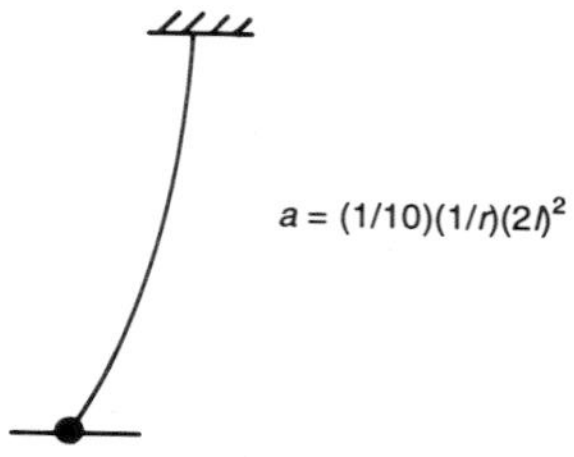

Fig. A.4. Column fixed at one end and pinned at the other

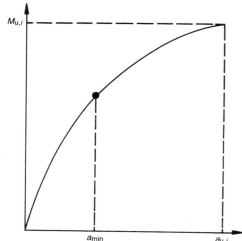

Fig. A.5. Moment–deflection curve for the critical column

Fig. A.6. Actual situation

Fig. A.5 shows the moment–deflection curve for the critical column. By definition, M_u corresponds to a_{min} for this column. In designing the other columns, reinforcement is selected to give the appropriate value of M_u: the design method does nothing to ensure that a deflection of a_{min} corresponds to this moment. The actual situation is likely to be as shown in Fig. A.6.

There is thus no compatibility between the required moment of resistance and the imposed deformation. Assuming the situation is as shown in Fig. A.6, a stronger column will have to be designed with a moment–deflection curve as shown in Fig. A.7.

In other words, a column should be designed such that the point (M_{des}, a_{min}) falls on its moment–deflection curve. In practice, the structure will be safe provided the curve either goes through or passes above the point (M_{des}, a_{min}). The problem is to find a simple way of defining an effective design value of M that will ensure this criterion is met.

APPENDIX: DESIGN OF SWAY FRAMES IN EC2

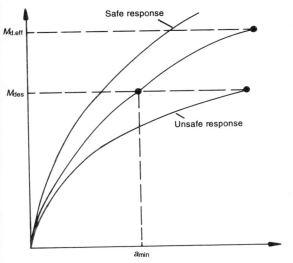

Fig. A.7. Moment–deflection curve of stronger column

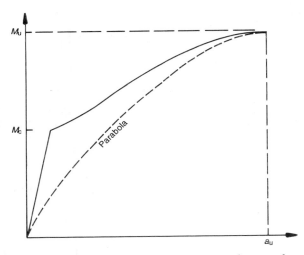

Fig. A.8. Moment–deflection curve for a column

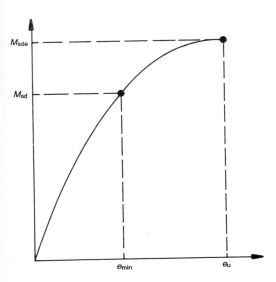

Fig. A.9. Parabolic moment–deformation curve

The moment–deflection curve for a column will be of the form shown in Fig. A.8.

It seems a reasonable and conservative approximation to assume a parabolic moment–deflection curve (the broken line in Fig. A.8). The design requirement can then be expressed as (see Fig. A.9)

$$\frac{M_{sd}}{M_{sde}} = \frac{e_{min}}{e_u}\left(2 - \frac{e_{min}}{e_u}\right)$$

An example is now used to explore the problem further and show how it may be solved in practice.

Example A.1
Design the columns in the structure shown in Fig. A.10.
Characteristic loads in columns.

Columns 1 and 6 $G_k = 540$ kN
                                     $Q_K = 270$ kN

All other columns $G_u = 1080$ kN
                                     $Q_u = 540$ kN

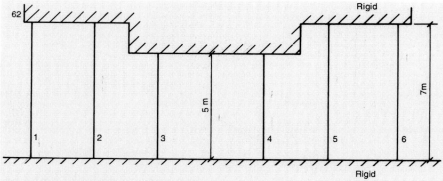

Fig. A.10. Example A.1 structure

For the purpose of the example it is assumed that the initial (first order) moments due to vertical loading in all the columns are zero, also, the effective length of the columns is assumed to be their actual length.

All columns are 300×300 mm and are made with 50 N/mm^2 concrete (cube) and 460 N/mm^2 reinforcement.

The shears carried by the columns will be a function of their relative stiffness. These are assessed to be as follows:

$$\begin{array}{ll} \text{Columns 3 and 4} & 18 \text{ kN} \\ \text{All other columns} & 6 \cdot 5 \text{ kN} \end{array}$$

Clearly columns 3 and 4 will be the most critical. EC2 equations 4.72 and 4.69 give:

$$e_2 = 44 \cdot 4 \, K_2 \text{ mm}$$

e_a will, for the sake of simplicity, be ignored.

(1) Imposed load + permanent loads only
 $N = 1 \cdot 35 \times 1080 + 1 \cdot 5 \times 540 = 2268$ kN
 $M = 0$
(2) Permanent loads + wind load
 $N = 1 \cdot 35 \times 1080 = 1485$ kN
 $M = H \, 1/2 = 1 \cdot 4 \times 18 \times 2 \cdot 5 = 63$ kN m
(3) Permanent loads + imposed load + wind load
 $N = 1 \cdot 35 \times 1080 + 1 \cdot 35 \times 540 = 2187$ kN
 $M = 1 \cdot 35 \times 18 \times 2 \cdot 5 \qquad = 60 \cdot 75$ kN m

It will be found that (3) gives the critical design condition for the column by a fairly wide margin.

From a design chart (see Figure 5.9), it can be found that:

$$K = 0 \cdot 44 \text{ hence ultimate deflection} = \cdot 44 \times 44 \cdot 4$$
$$= \underline{19 \cdot 6 \text{ mm}}$$

hence
$$\begin{aligned} N_{sd} &= 2187 \text{ kN} \\ M_{sd} &= 60 \cdot 75 + 2187 \times \cdot 0196 \\ &= 103 \cdot 6 \text{ kN m} \end{aligned}$$

lReinforcement ratio = $\underline{3 \cdot 1\%}$

The deflection of the remaining columns will be $19 \cdot 6$ mm under the condition leading to failure of columns 3 and 4.

Column 2
Considering only the case of dead and imposed + wind loads,

$$\begin{aligned} N &= 2187 \text{ kN (as column 3)} \\ M_i &= 1 \cdot 35 \times 6 \cdot 5 \times 2 \cdot 5 = 30 \cdot 7 \text{ kN m} \\ M_2 &= 2187 \times 19 \cdot 6/1000 = 42 \cdot 86 \text{ kN m} \\ M_{tot} &= 73 \cdot 57 \text{ kN m} \end{aligned}$$

From the design chart, this gives a steel percentage of 2·3% and a value of K of 0·35.

The ultimate deflection implied by a value of K of 0·35 is 30 mm. This is very different to the 19·6 mm which was assumed hence, at a deflection of 19·6, the resisting moment will be significantly less than 63·79 kN m and equilibrium will not be satisfied.

Using the device proposed earlier, we need to find a moment which satisfies the relation:

$$\frac{73\cdot57}{M_{sde}} = \frac{19\cdot6}{e_u}\left(2 - \frac{19\cdot6}{e_u 0}\right)$$

Since $e_u = 87\cdot1\, K_2$ and since the design chart uses the variable M/bh^2, it is convenient to re-arrange the equation as follows:

$$2\cdot72/(M_{sde/bh^2}) = \frac{\cdot225}{K_2}\left(2 - \frac{\cdot225}{K_2}\right)$$

From the design chart, we can establish the relationship between K_2 and M/bh^2 for an axial stress of $2187 \times 10^3/300^2 = 24\cdot3$ and hence find a design which satisfies the above relationship. This is, roughly

$$M_{sde}/bh^2 = 3\cdot4 \text{ hence } M_{sde} = 92 \text{ kN m}$$
$$K_2 = 0\cdot405$$
$$100\, A_s/bh = 2\cdot8\%$$

Thus, it has been necessary in this case to increase the moments by 26% above those estimated simply on the basis of the ultimate deflection.

The procedure for establishing an appropriate value of M/bh^2 can be illustrated graphically as shown below though, in practice, it will not normally be necessary to plot the graph.

Taking a horizontal line across the design chart corresponding to the design axial load, it is easy to plot K_2 against M/bh^2. Likewise, the equation

$$\frac{2\cdot72}{(M_{sde}/bh^2)} = \frac{\cdot225}{K_2}\left(2 - \frac{\cdot225}{K_2}\right)$$

can also be plotted. The correct value of M/bh^2 is given by the point at which the two curves intersect (see Figure 12).

Edge columns

The design axial load and initial moment on the edge columns are respectively

$$N_{sd} = 1093 \text{ kN}$$
$$M_i = 30\cdot7 \text{ kN m}$$

For a deflection of 19·6 mm, the additional moment due to second-order effects is 21·4 kN m, hence the total design moment = 43·3 kN m. Using the proposed relationship, the condition

$$\frac{52\cdot1}{M_{sde}} = \frac{19\cdot6}{e_u}\left(2 - \frac{19\cdot6}{e_u}\right)$$

must be satisfied. As before, this can be conveniently simplified to

$$\frac{1\cdot93}{M_{sde}/bh^2} = \frac{0\cdot225}{K_2}\left(2 - \frac{0\cdot225}{K_2}\right)$$

The design chart gives the required value of M/bh^2 of 4·3, a moment of 118 kN m and a steel percentage of 1·20%. This is a very substantial increase in moment (more than double).

A further exercise may, however, be of interest. Assume that *all* the columns are 7 m long and find the steel areas required.

In this cae, for an internal column

$$N_{sd} = 2187 \text{ kN}$$

$$\text{Horizontal load} = 1 \cdot 35 \times 62/6$$
$$= 14 \text{ kN}$$
$$\text{Hence, initial moment} = 14 \times 3 \cdot 5$$
$$= 49 \text{ kN m}$$
$$\text{Additional eccentricity} = 250 \times 3484 \times K_2$$
$$= 87 \cdot 1 \, k_2$$
$$\text{Hence } M_2 = 190 \cdot 5 \, K_2$$

From the design chart, this gives:

$$K_2 = 0 \cdot 53 \quad M = 150 \text{ kN m} \quad 100 \, A_s/bh = 4 \cdot 5\%$$

The design for the edge column gives $K_2 = 0 \cdot 9$, $M = 135$ kN m and $100 \, A_s/bh = 1 \cdot 8\%$. Both these are greater than the percentages calculated above, which would be expected, so, intuitively, the result from the proposed method seems reasonable.

BS8110 proposes that the average value of e_2, calculated for each column ignoring the other columns should be taken for a sway structure. This approach would have given the following ultimate deflections.

$$\text{columns 3 \& 4} \quad 19 \cdot 6 \text{ mm}$$
$$\text{columns 2 \& 5} \quad 43 \cdot 6 \text{ mm}$$
$$\text{columns 1 \& 6} \quad \underline{79 \cdot 5 \text{ mm}}$$

$$\text{mean } e_2 = 47 \text{ mm}$$

This would have given the following designs for the various columns by the BS8110 method compared with the proposal as:

BS8110		Proposal	
Moment	$100 \, A_s/bh$	Moment	$100 \, A_s/bh$
164	4·9	104	3·1
134	4·0	92	2·8
82	0·4	118	1·8

It is worth noting that there is no unique solution to the sway frame problem and alternative solutions to the result presented could be devised. The principle behind the method is the definition of an ultimate deflection. In the example, using normal design procedures, an ultimate deflection was found for the most critical column. This, in fact, is a minimum value for the ultimate deflection. If more reinforcement had been put into the critical column, its ultimate deflection would have been greater. This can be seen from the design chart where, for axial loads above the balance, K increases with increasing steel percentage. The effect of increasing the reinforcement, and hence the deflection of the most critical column, leads to a redistribution of the lateral forces on the columns.

Recourse to the example will clarify this.

Suppose it was decided to put 5% of reinforcement in columns 3 and 4. From the design chart, this corresponds to an ultimate moment of 170 kN m. The value of K_2 is 0·57 and hence the deflection at ultimate is 25·3 mm.

The second order moment is thus $0 \cdot 0253 \times 2186 = 55 \cdot 3$ kN m. Hence, the column is supporting a first order moment of $170 - 55 \cdot 3 = 114 \cdot 7$ kN m rather than the moment of 60·75 originally calculated. This corresponds to a lateral load of $114 \cdot 7/2 \cdot 5 = 45 \cdot 9$ kN. Clearly then, the lateral loads carried by the other columns must be reduced. The total design lateral force $= 62 \times 1 \cdot 35 = 83 \cdot 7$ kN. The two internal columns are each supporting 45·9 kN hence the shear carried by the other columns is reduced to $(83 \cdot 7 - 2 \times 45 \cdot 9)$. This is -2 kN per column compared with an initial calculated value of 8·8.

The conditions in columns 2 and 5 are thus now

$$N_{sd} = 2186$$
$$M_i = -7$$
$$M_2 = 25\cdot3 \times 2186/1000 = 55\cdot3$$

Total moment = 48·3 kN m.

applying the method described in this paper we can write

$$\frac{48\cdot3}{M_{sde}} + \frac{25\cdot3}{e_u}\left(2 - \frac{25\cdot3}{e_u}\right)$$

hence:

$$\frac{1\cdot79}{(M_{sde}/bd^2)} = \frac{\cdot29}{K_2}\left(2 - \frac{\cdot29}{K_2}\right)$$

This gives $M_{sde} = 49$, $K = 0\cdot26$, $100\,A_s/bh = 1\cdot75$
This is a reduction on the previous value of 2·8%.

It was argued earlier that, having defined the ultimate deflected shape of the most critical column, a design had to be devised for all the other columns so that they were each in equilibrium with the applied loads when this deflection was applied. This basic principle can be generalised in a way which will indicate how further flexibility can be given to the designer.

The requirement that under the critical deformation, which is that deformation which will lead to failure of the first column, the whole structure must be capable of sustaining the applied external loads. This is illustrated in Figure 13, which is a generalised version of Figure 7.

The actual design load-deformation characteristic must either pass through the point X or pass above it. Any design system that ensures that this is achieved will produce safe designs. In the arguments developed earlier in this note, this has been achieved by ensuring that each column in turn is capable of withstanding the loads applied to it with the defined deflection. In the first section of the example, the loading carried by each column was defined a-prior by elastic analysis. In the final example, the lateral loads in the columns were arrived at as a result of choosing a more arbitrary design for the critical column, establishing the lateral load it supported in its ultimate condition and then sharing the remainder out among the other columns (in the case of the example, the share carried by the other columns was negative).

This second idea can be carried further. If there are columns whose ultimate deflections are very high compared with the critical deflection, it will require a large increase in their reinforcement area to provide the required lateral moment capacity. In such cases, it is permissible to ignore their lateral load capacity and not include them in the sway analysis. This will be conservative since they will actually carry some lateral load under the deflection considered.

Again, this can be explored by means of the example. Assume that we wish the lateral load capacity of the two edge columns to be ignored. This revises the loads carried by the columns to:

Columns	Vertical load (ult)	Lateral load (ult)
1 & 6	1093	0
2 & 5	2187	11·2
3 & 4	2187	30·6

Designing columns 3 & 4 by thge standard method gives:

$$M_{sd} = 125\cdot2$$
$$K = 0\cdot5, \text{ hence } a_{min} = 22\cdot2 \text{ mm}$$
$$100\,A_s/bh = 3\cdot8\%$$

Columns 2 and 5 now have to support a moment of $22\cdot2 \times 2187/1000 = 48\cdot6$ kN m, giving a total moment of $48\cdot6 + 11\cdot2 \times 3\cdot5 = 87\cdot8$.

Using the proposal in this paper,

$$\frac{87 \cdot 8}{M_{sde}} = \frac{22 \cdot 2}{e_u}\left(2 - \frac{22 \cdot 2}{e_u}\right)$$

This leads to:

$$M_{sde} = 108 \text{ N m}$$

and $100 \, A_s/bh = 3 \cdot 2\%$

The edge columns are designed for a vertical load of 1093 kN at an eccentricity. This only requires nominal reinforcement.

Study of these examples shows that a high degree of flexibility is available in the design of sway frames. This can be seen from the following table which compares the design moments in the three cases considered.

Columns	Case 1 (direct design)	Case 2 (increase a_{min})	Case 3 (ignore outer columns)
1 & 6	118	27	24
2 & 5	92	49	108
3 & 4	104	170	125

The relative economy is probably better seen from the following table which compares the steel percentages in the columns. In comprising the figures, it should be noted that EC2 Equation 5.13 gives a minimum steel percentage for the edge columns of 0·47%. This governs in two cases, though the area actually required to resist the design axial load and moment is zero.

It should be noted that, in all cases, it is also necessary to check each individual column for deformation as an isolated column. To avoid complicating the presentation of the argument, this has not been done but it would not have resulted in any change in the designs.

Columns	Case 1 (direct design)	Case 2 (increase e_{min})	Case 3 (ignore outer columns)
1 & 6	1·8	0·47	0·47
2 & 5	2·8	1·75	3·20
3 & 4	3·1	5·00	3·80
Total	7·10	7·15	7·40

It appears that the total weight of steel required in the columns is only marginally affected by the approach adopted.

Summary of design method

We can now begin to drawn this note together by summarising the design method that has resulted from the theoretical consideration and the example. This can be done as follows:

In designing sway frames the following basic assumptions are made.

(1) calculate, using K_2 factor charts, the ultimate eccentricity for each column, subjected to its design vertical and horizontal loads and moments.
(2) identify the smallest ultimate eccentricity e_{min}.
(3) for all other columns, establish a design moment from the K_2 factor charts which will satisfy the relationship:

$$\frac{M_{sd}}{M_{sde}} = \frac{e_{min}}{K_2 e_{v,i}}\left(2 - \frac{e_{min}}{K_2 e_{v,i}}\right)$$

where e_{min} is the critical deflection
$e_{v,i}$ is the ultimate eccentricity, calculated assuming $K_2 = 0$ for the ith column
M_{sd} is the design moment $= M_o + N_{sd} e_{min}$
M_{sde} is the effective design ultimate moment which should be used in establishing the required reinforcement.

(4) Design each column to withstand:
$$N_{sd} + M_{sde}$$
(for the critical column, $M_{sde} = M_{sd}$).

(5) It will be conservative, and occasionally more economical to ignore the lateral load carrying capacity of some columns. From the example, it would seem reasonable to ignore columns where the ultimate deflection of the column, calculated ignoring interaction with the rest of the structure, exceeds $3 e_{min}$. When this is done, steps (1)–(4) are carried out for the remaining columns. The columns which have been neglected in the analysis are then designed for their design axial load plus a moment equal to $N_{sd} e_{min}$.

Concluding discussion

There are two basic issues which, in considering this approach, should be considered separately. These are:

(a) the basic theory and
(b) simplified rules.

If the theory is wrong, then the effectiveness of the rules is irrelevant. However, if the theory is right but the rules wrong, then the rules can be modified or improved. It may, in any case, be possible to develop rules which are less clumsy to operate than those proposed.

To make assessment easier, the basic theory is as follows.

Assumptions

The following assumptions apply.

(a) All columns in a given storey will deform by the same amount.
(b) Failure of the structure occurs when the first column reaches its ultimate deformed shape (critical deformation).

Theory

Under the critical deformation, the whole structure must be capable of sustaining the applied external loads. This can be expressed as a requirement that the design load and critical deformation specify a point which either lies on or above the load-deformation characteristic of the total structures (see Fig. 13).

Thus
$$S_d \le S_{(e_{crit})}$$

where $S_{e_{crit}}$ is the loading corresponding to the critical deformation.

A legitimate way of ensuring that the inequality above is satisfied is to ensure that it is satisfied for each column individually. If this is done then the total structure must satisfy the inequality.

Rules

(a) It is assumed that the curvature of a column at failure is given by Eq. 4.72 in EC2. This is
$$1/r = \frac{\cdot 8 \, K_2 \, f_{yd} \, d}{E_s}$$

where $K_2 = (N_{ud} - N_{sd})/(N_{ud} - N_{bai}) \le 1$

(b) The second order eccentricity of a column is given by Eq. 4.69 in EC2. This is

$$e_2 = 0.1 K_1 l_o^2 (1/r)$$

K_1 is given by $/20 - 0.75$ for $15 \leq \leq 35$ ($= 35$ corresponds to l_o/h so for most practical situations, $K_1 = 1$).

l_o is the effective length of the column.

(c) The moment–deformation relation for a column may be conservatively assumed to be a parabola with its peak at e_2 calculated for that particular column considered in isolation.

All the rules are clearly simplified models of behaviour which could be improved upon.